Handbook of Sludge-Handling Processes

Water Management Series
Series Editor: Gordon L. Culp

**Handbook of Wastewater Collection and Treatment:
Principles and Practice**
By: M. Anis Al-Layla
Shamim Ahmad
E. Joe Middlebrooks

**Handbook of Biological Wastewater Treatment:
Evaluation, Performance, and Cost**
By: Henry H. Benjes, Jr.

**Handbook of Sludge-Handling Processes:
Cost and Performance**
By: Gordon L. Culp

Handbook of Sludge-Handling Processes

Cost and Performance

Gordon L. Culp

Water Management Series

Garland STPM Press/New York & London

Copyright © 1979 by Garland Publishing, Inc.

All rights reserved. No part of this work covered by the copyright hereon may be reproduced or used in any form or by any means—graphic, electronic, or mechanical, including photocopying, recording, taping, or information storage and retrieval systems—without permission of the publisher.

15 14 13 12 11 10 9 8 7 6 5 4 3 2 1

Library of Congress Cataloging in Publication Data

Culp, Gordon L.
  Handbook of sludge handling processes.

  (Water management series)
  Bibliography: p.
  Includes index.
  1. Sewage sludge. I. Title. II. Series.
TD767.C84   628'.36   78-20644
ISBN 0-8240-7062-3

Published by Garland STPM Press
545 Madison Avenue/New York, New York 10022

Printed in the United States of America

Water Management Series

The Garland series in water management addresses the many aspects of water resource management which must be carefully integrated to protect the quality and quantity of our water supplies. The series is a comprehensive compilation of volumes dealing with the most important aspects of water resource management. Topics span the range from detailed information on design, cost, and performance of specific types of treatment processes to the environmental implications associated with certain management approaches. The series is designed to be a timely and authoritative reference source for practicing engineers and planners, as well as for regulatory agencies and students.

Series Editor:
Gordon L. Culp

Contents

Preface — vii

1
The Significance of Sludge Management — 1

2
Sludge Conditioning — 11

3
Sludge Thickening, Pumping, and Storage — 45

4
Sludge Dewatering — 71

5
Sludge Stabilization — 113

6
Sludge Incineration and Drying — 149

7
Disposal and Land Application — 183

8
Sludge Transport — 197

References — 217

Index — 223

Preface

The successful handling of sludge from wastewater-treatment plants has long been a difficult task. Widespread use of higher levels of wastewater treatment has resulted in even more difficult and costly situations for many municipalities and industries. It is important that all alternatives which exist for handling sludges at a given locale be carefully evaluated before selection of a sludge-management plan. The alternatives must be evaluated in terms of their technical feasibility, energy consumption (or, in some cases, production), environmental impacts, and costs. The literature contains many articles, texts, and manuals which describe the theory, design, and operation of alternative sludge-handling processes. This text does not compile and repeat this information, but rather is intended to fill what the author perceives as a void in the published literature. There is a lack of a single source of practical, realistic information on the performance, energy aspects, and cost of alternative sludge-handling processes. The ready availability of such information would be very valuable to the engineer and/or planner in evaluating alternative sludge-management plans for a locale, and in evaluating specific, alternative processes; to the municipal official and concerned public in determining the reasonableness of a proposed or current plan; and to the student who desires to supplement his study of the theoretical aspects of process design with a study of factors which govern the real-world acceptability of a process. Thus, this text concentrates on presenting data on performance, energy, and costs of alternative sludge processes with no attempt to serve as a review of theory or as a design manual.

The reader is cautioned about applying the results and costs cited for one locale to another without careful consideration of all the factors involved. An important factor to consider is the variability of sludge characteristics between geographical areas and often within the same system. For example, two wastewater-treatment systems in different locations may use identical processes, but the sludge produced can vary in nature. Seasonal factors, such as industrial waste loads from a cannery, can cause variations in a given system. Because of variability in characteristics, there is no universal solution to treating and disposing of sludges.

The economics change from location to location, as does the environmental acceptance of alternative processes.

Much of the cost and energy information presented in this text have been gathered and analyzed by Culp/Wesner/Culp during the course of several contracts with the U.S. Environmental Protection Agency, especially contracts 68–03–2186 and 68–03–2516 from the Municipal Environmental Research Laboratory in Cincinnati, Ohio. The substantial contribution of these contracts is gratefully acknowledged, as are those of the following specific individuals within Culp/Wesner/Culp who have participated in these efforts: George M. Wesner, Russell L. Culp, William F. Ettlich, Henry H. Benjes, Jr., Daniel J. Hinrichs, Sigurd P. Hansen, Robert C. Gumerman, Robert B. Williams, Nancy E. Folks-Heim, and Thomas E. Lineck. The contributions of performance data and cost information, and the time and patience of the individuals from many cities, industries, regulatory agencies, and equipment manufacturers is sincerely appreciated and acknowledged.

Gordon L. Culp
May 1978

Chapter 1

The Significance of Sludge Management

SLUDGE QUANTITIES

Implementation of wastewater-management plans which result in higher degrees of wastewater treatment will produce improvements in water quality. Unfortunately, these improvements will be accompanied by the production of increasing quantities of increasingly difficult-to-handle sludges. For example, primary treatment of municipal wastewaters typically produces 2,500–3,000 gallons of sludge per million gallons of wastewater treated. When treatment is upgraded to activated sludge, the sludge quantities increase by 15,000–20,000 gallons per million gallons treated. Use of chemicals for phosphorus removal can add another 10,000 gallons per million gallons. The sludges withdrawn from primary treatment are as much as 97% water. Secondary and many chemical sludges have higher water contents and are much more difficult to dewater than primary sludges. A recent projection of sludge trends in the U.S.[1] indicates the magnitude of the problem, shown in Table 1–1.

COSTS

The cost of sludge handling and disposal is often greater than the cost of treating the wastewater itself. For example, the cost (capital, operation, and maintenance) of providing secondary treatment of 10 million gallons per day of municipal wastewater may be 20–25 cents per 1,000 gallons, while the cost of disposing of the resulting sludges may be 30%—100% (or more) of this amount. Another significant consideration is that although there may be several environmentally acceptable, technically feasible, and economically competitive methods for wastewater treatment in a given area, there may be only a few—perhaps only one—such sludge-disposal alternatives. Thus, sludge-disposal considerations are an important element in the selection of an overall wastewater-management plan.

Regionalization of several smaller wastewater systems into a larger

TABLE 1-1
Sludge Trends in the U.S.

Sludge Type	1972		1985	
	Population (million)	Tons/Year (million)	Population (million)	Tons/Year (million)
Primary	145	3.20	170	3.7
Secondary	101	1.50	170	2.5
Chemical	10	0.09	50	0.5
Total		4.8		6.7

system may favorably affect the relative economics of some sludge disposal alternatives to the degree that they become economically feasible, whereas they were not for any of the individual, smaller plants. For example, heat drying of sludge for use as a fertilizer decreases in cost by a factor of nearly two as plant capacity increases from 10 to 50 million gallons per day, while the cost of the anaerobic digestion process decreases only about 10%.

BASIC DISPOSAL ALTERNATIVES

Although a large number of alternative combinations of equipment and processes are used for treating sludges, the basic alternatives are fairly limited.[2] The ultimate depository of the materials contained in the sludge must either be land, air (by-products of incineration), or water. Current policies discourage practices such as ocean dumping of sludge as long-term solutions. Air-pollution considerations necessitate air-pollution facilities as part of the sludge-incineration process. Incineration results in a residual ash which must be disposed of. Thus, sludge in some form will eventually be returned to the land.

There are two basically different philosophical approaches in handling the sludges from wastewater treatment: reuse and disposal. The reuse approach is based on recycling the sludges so that nutrients and organics contained in the sludges are beneficially reused. The goal of sludge treatment in this case is to make the sludge compatible with the proposed reuse system (i.e., to stabilize the sludge so that it will not cause nuisance conditions, will eliminate pathogens to prevent disease problems, etc.). The organic solids which make up 60%–80% of the solids in a typical municipal sludge also are a potential source of energy (typically about

10,500 BTUs per pound). Some processes convert these solids so that this energy can be beneficially reused.

The disposal philosophy considers the sludge a waste material. In some cases, such as ocean dumping, limited pretreatment of sludge is provided prior to disposal. Most disposal systems, however, incorporate treatment techniques to provide maximum reductions in sludge volume prior to disposal with little or no regard for the potentially beneficial components of sludge.

The choice between the disposal and reuse approaches must be based on an evaluation of the many factors (economics, environmental impacts, energy consumption, etc.) associated with each of the processes involved. For example, the feasibility of recycling sludges to the land for agricultural reuse is dependent on the quality of the sludge and the availability, location, use, nature, and cost of land. These factors may be a problem in some urban areas. Chicago transports a portion of its sludges to a site 160 miles downstate for application to previously strip-mined land, indicating that under some circumstances transport to even relatively distant locales is practical.

In some cases, integration of solid-waste and wastewater-sludge disposal plans may offer a useful, synergistic relationship. For a given population, the volume of solid wastes is about 10 times the volume of wastewater sludge. Thus, inclusion of sludge with solid wastes from an area may not significantly alter the volume of material to be handled in the solid-waste system. For example, under certain conditions disposal of dewatered, stabilized sludge may be readily compatible with an existing solid-waste landfill practice. On the other hand, a sludge-handling system would have to be altered drastically to have capacity for solid wastes. Such a plan could unfavorably affect an otherwise acceptable sludge-handling plan. For example, a local market for compost might be ample for composted sewage sludge but might be overwhelmed by the much larger volume of a composting operation handling both sludge and solid wastes.

Figure 1-1 summarizes general sludge-handling alternatives. Recent Environmental Protection Agency (EPA) reports[3,4] summarize the current use of these various alternatives in the U.S., with a breakdown for disposal of municipal sewage sludge on a national basis, as shown in Table 1-2.

SLUDGE CHARACTERISTICS

The composition of sewage sludges from municipal systems varies widely from one locale to another depending on a variety of factors. The presence

4/Handbook of Sludge-Handling Processes

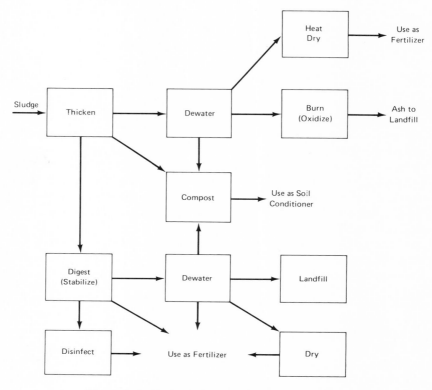

Figure 1-1. Basic sludge-handling alternatives.

TABLE 1-2
Nationwide Disposal of Municipal Sewage Sludge

Method	% Total municipal sludge
Ocean Disposal	15
Incineration	35
Landfill	25
Land Application	
croplands	20
other	5

The Significance of Sludge Management/5

or absence of industrial wastes can have a profound effect on the quantity and quality of wastewater sludges. The chemical quality of the community's raw water supply will affect the chemical composition of the wastewater sludges. The presence or absence of stormwater in the system will also affect the sludge composition (i.e., by the amount of grit carried into the plant). Table 1-3 summarizes typical components of sewage sludges in 150 treatment plants in the north-central and eastern U.S.

Because the nature of sludges resulting from the treatment of municipal wastewaters varies so greatly from one locale to another, generalized statements about their nature are of limited value. But here are some observations which, however, are generally true.

Raw primary sludges almost universally settle, thicken, and dewater with relative ease compared to secondary biological sludges because of their coarse, fibrous nature. Generally, at least 30% of the solids are larger than 30 mesh in size. These coarse particles permit rapid formation of a sludge cake with sufficient structural matrix to permit good solids capture and rapid dewatering. Contrary to some reports in the literature, anaerobic digestion of primary sludges frequently makes them somewhat more difficult to thicken and dewater. The dewatering results, however, are still generally good at relatively low cost.

Activated sludges show much greater variation in dewatering characteristics than do primary sludges. These variations may even be substantial from day to day at the same plant. The sludges are much finer than primary sludges and are largely cellular organic material with a density very nearly the same as water. They are much more difficult to dewater than primary sludges.

The nature of sludges resulting from the chemical coagulation of sewage depends on the nature of the coagulant used. Generally, alum and iron coagulants produce gelatinous floc which is difficult to dewater. Lime coagulation produces a sludge which readily thickens and dewaters in most cases. Estimates of sludge quantities and characteristics from a variety of wastewater processes may be found in the EPA's design manual for sludge treatment and disposal.[6]

OPERATIONS UNIT FOR SLUDGE MANAGEMENT

As noted in the preface, it is not the purpose of this text to provide detailed design guidance. Other readily available references[6-10] provide such information. The subsequent chapters briefly describe available sludge-processing and management alternatives, and then discuss as-

TABLE 1-3
Major Components of Sludge

Component	Sample Type	Number	Range	Median	Mean
Organic C (%)	Anaerobic	31	18–39	26.8	27.6
	Aerobic	10	27–37	29.5	31.7
	Other	60	6.5–48	32.5	32.6
	All	101	6.5–48	30.4	31.0
Total N (%)	Anaerobic	85	0.5–17.6	4.2	5.0
	Aerobic	38	0.5–7.6	4.8	4.9
	Other	68	<0.1–10.0	1.8	1.9
	All	191	<0.1–17.6	3.3	3.9
NH_4-N (ppm)	Anaerobic	67	120–67,000	1,600	9,400
	Aerobic	33	30–11,300	400	950
	Other	3	5–12,500	80	4,200
	All	103	5–67,600	920	6,540
NO_3-N (ppm)	Anaerobic	35	2–4,900	79	520
	Aerobic	8	7–830	180	300
	Other	3	—	—	780
	All	45	2–4,900	140	490
Total P (%)	Anaerobic	86	0.5–14.3	3.0	3.3
	Aerobic	38	1.1–5.5	2.7	2.9
	Other	65	<0.1–3.3	1.0	1.3
	All	189	<0.1–14.3	2.3	2.5
Total S (%)	Anaerobic	19	0.8–1.5	1.1	1.2
	Aerobic	9	0.6–1.1	0.8	0.8
	Other	—	—	—	—
	All	28	0.6–1.5	1.1	1.1
K (%)	Anaerobic	86	0.02–2.64	0.30	0.52
	Aerobic	37	0.08–1.10	0.38	0.46
	Other	69	0.02–0.87	0.17	0.20
	All	192	0.02–2.64	0.30	0.40
Na (%)	Anaerobic	73	0.01–2.19	0.73	0.70
	Aerobic	36	0.03–3.07	0.77	1.11
	Other	67	0.01–0.96	0.11	0.13
	All	176	0.01–3.07	0.24	0.57
Ca (%)	Anaerobic	87	1.9–20.0	4.9	5.8
	Aerobic	37	0.6–13.5	3.0	3.3
	Other	69	0.1–25.0	3.4	4.6
	All	193	0.1–25.0	3.9	4.9

TABLE 1-3 (continued)

Component	Sample Type	Number	Range	Median	Mean
Mg (%)	Anaerobic	87	0.03–1.92	0.48	0.58
	Aerobic	37	0.03–1.10	0.41	0.52
	Other	65	0.03–1.97	0.43	0.50
	All	189	0.03–1.97	0.45	0.54
Ba (%)	Anaerobic	27	<0.01–0.90	0.05	0.08
	Aerobic	10	<0.01–0.03	0.02	0.02
	Other	23	<0.01–0.44	<0.01	0.04
	All	60	<0.01–0.90	0.02	0.06
Fe (%)	Anaerobic	96	0.1–15.3	1.2	1.6
	Aerobic	38	0.1–4.0	1.0	1.1
	Other	31	<0.1–4.2	0.1	0.8
	All	165	<0.1–15.3	1.1	1.3
Al (%)	Anaerobic	73	0.1–13.5	0.5	1.7
	Aerobic	37	0.1–2.3	0.4	0.7
	Other	23	0.1–2.6	0.1	0.3
	All	133	0.1–13.5	0.4	1.2
Mn (mg/kg)	Anaerobic	81	58–7,100	280	400
	Aerobic	38	55–1,120	340	420
	Other	24	18–1,840	118	250
	All	143	18–7,100	260	380
B (mg/kg)	Anaerobic	62	12–760	36	97
	Aerobic	29	17–74	33	40
	Other	18	4–700	16	69
	All	109	4–760	33	77
As (mg/kg)	Anaerobic	3	10–230	116	119
	Aerobic	—	—	—	—
	Other	7	6–18	9	11
	All	10	6–230	10	43
Co (mg/kg)	Anaerobic	4	3–18	7.0	8.8
	Aerobic	—	—	—	—
	Other	9	1–11	4.0	4.3
	All	13	1–18	4.0	5.3
Mo (mg/kg)	Anaerobic	9	24–30	30	29
	Aerobic	3	30–30	30	30
	Other	17	5–39	30	27
	All	29	5–39	30	28

TABLE 1-3 (*continued*)

Component	Sample Type	Number	Range	Median	Mean
Hg (mg/kg)	Anaerobic	35	0.5–10,600	5	1,100
	Aerobic	20	1.0–22	5	7
	Other	23	2.0–5,300	3	810
	All	78	0.5–10,600	5	733
Pb (mg/kg)	Anaerobic	98	58–19,730	540	1,640
	Aerobic	57	13–15,000	300	720
	Other	34	72–12,400	620	1,630
	All	189	13–19,700	500	1,360
Zn (mg/kg)	Anaerobic	108	108–27,800	1,890	3,380
	Aerobic	58	108–14,900	1,800	2,170
	Other	42	101–15,100	1,100	2,140
	All	208	101–27,800	1,740	2,790
Cu (mg/kg)	Anaerobic	108	85–10,100	1,000	1,420
	Aerobic	58	85–2,900	970	940
	Other	39	84–10,400	390	1,020
	All	205	84–10,400	850	1,210
Ni (mg/kg)	Anaerobic	85	2–3,520	85	400
	Aerobic	46	2–1,700	31	150
	Other	34	15–2,800	118	360
	All	165	2–3,520	82	320
Cd (mg/kg)	Anaerobic	98	3–3,410	16	106
	Aerobic	57	5–2,170	16	135
	Other	34	4–520	14	70
	All	189	3–3,410	16	110
Cr (mg/kg)	Anaerobic	94	24–28,850	1,350	2,070
	Aerobic	53	10–13,600	260	1,270
	Other	33	22–99,000	640	6,390
	All	180	10–99,000	890	2,620

pects such as results, energy aspects, and costs. Unit operations of a similar function are grouped in a single chapter, such as conditioning, thickening, dewatering, stabilization, incineration, drying, land application, and transport.

Whenever possible, operating and maintenance costs are expressed in terms of hours of labor, kilowatt-hours of energy, or other units which can be multiplied by appropriate local unit costs. Capital costs for the more commonly used unit processes, in addition to being presented in a summary curve, are tabulated by cost component to enable the costs presented to be updated and improved as more cost data are available and as inflation affects costs. Cost components presented include the following:

Manufactured Equipment: This item includes estimated purchase cost of pumps, drives, process equipment, and other items which are factory made and sold as equipment.

Concrete Products: This item includes the delivered cost of ready-mix concrete.

Steel: This item includes reinforcing steel for concrete.

Miscellaneous Steel: This item includes steel for weirs and launders, but does not include steel pipe or structural steel for housing.

Metal Pipe and Valves: Cast-iron pipe, steel pipe, valves, and fittings have been combined into a single component. This item includes the purchase price of pipes, valves, fittings, and support devices.

Housing: In lieu of segregating building costs into its several components, this category represents all material and labor costs associated with the building, including heating, ventilating, air conditioning, and lighting.

Electrical and Instrumentation: The cost of electrical service and instrumentation associated with the process equipment is included in this item.

Special Construction: The provision for cost items peculiar to a specific unit process are provided in this classification.

Labor: The labor associated with excavation, installing equipment, piping, and valves, constructing concrete forms, and placing concrete and reinforcing steel are included in this item.

Miscellaneous Items: This provides a category for those items which have not been quantified, but which, with a detailed quantity takeoff, would be defined. These items generally represent 10%–20% of the total cost of the facility and include such things as handrails, access hatches, portable and stationary miscellaneous small equipment, and contractor's costs associated with equipment use.

Contingency: The contingency item is intended to represent costs which develop in the facility as the design requirements become better

TABLE 1-4
BLS Price Indices for Cost Components of Sludge Handling

	Price category	Index number
Equipment	BLS general purpose machinery & equipment	114
Concrete	BLS concrete ingredients	132
Steel	BLS steel mill products	1013
Labor	ENR wage index (skilled labor)	—
Electrical & instrumentation	BLS electrical machinery & equipment	117

defined, and also provides a cushion in the event inflation exceeds the anticipated rate.

The appropriate BLS wholesale price index for each of these categories is shown in Table 1-4.

Where it has not been practical to break down a system into individual components, the EPA sewage-treatment cost indices may be applied to approximate the effects of inflation.

Chapter 2
Sludge Conditioning

The purpose of sludge conditioning is to increase the rate and/or extent of dewatering achievable for a given sludge. A wide variety of physical and chemical techniques is used. The use of sludge conditioning prior to dewatering has become standard practice, and their combined result permits the moisture content of sludges to be reduced by subsequent dewatering from 95%–98% to 60%–75%.

CHEMICAL CONDITIONING

The most frequently encountered conditioning practice in the U.S. today is the use of ferric chloride, either alone or in combination with lime,[7] although the use of polymers is rapidly gaining widespread acceptance. While ferric chloride and lime are normally used in combination, it is not unusual for them to be applied individually. Lime alone is a fairly popular conditioner for raw primary sludge and ferric chloride alone has been used for conditioning activated sludges. Lime treatment to a high pH value has the added advantage of providing a significant degree of disinfection of the sludge.[11]

Organic polymeric coagulants and coagulant aids that have been developed in the past 20 years are rapidly gaining acceptance for sludge conditioning.[12] These polyelectrolytes are of three basic types:

1. *Anionic* (negative charge)—serve as coagulant aids complementing inorganic Al^{3+} and Fe^{3+} coagulants by increasing the rate of flocculation, size, and toughness of particles.
2. *Cationic* (positive charge)—serve as primary coagulants or in conjunction with inorganic coagulants.
3. *Nonionic* (equal amounts of positively and negatively charged groups in monomers)—serve as coagulant aids in a manner similar to that of both anionic and cationic polyelectrolytes.

The popularity of polymers is primarily due to their ease in handling, small storage space requirements, and effectiveness. All of the inorganic coagulants are difficult to handle, and their corrosive nature can cause maintenance problems in the storing, handling, and feeding systems in addition to the safety hazards inherent in their handling. Many plants in the U.S. have abandoned the use of inorganic coagulants in favor of polymers.

The facilities for chemical conditioning are relatively simple and consist of equipment to store the chemical(s), feed the chemical(s) at controlled dosages, and mix the chemical(s) with the sludge. The cost of chemical conditioning is primarily a function of the quantity of chemical required, which is affected by factors such as:

1. Solids concentration.
2. Sludge particle size.
3. Proportion of volatile matter in sludge.
4. Reducing agents in sludge, i.e., H_2S.
5. Alkalinity.

The chemical requirements for any given sludge can be determined accurately only by tests on the specific sludge involved. Typical values are shown in Table 2-1.

TABLE 2-1
Chemical Requirements for Sludge Conditioning[a]

	$FeCl_3$ (lbs/ton dry solids)	Lime (lbs CaO/ton dry solids)	Polymer (lbs/ton dry solids)
Raw primary + activated sludge	40–50	200	15–20
Digested primary + activated sludge	80–100	160–300	30–40
Elutriated primary + activated sludge	40–125	—	20–30

[a]*Note:* The use of the inorganic chemical conditioning chemicals can increase the weight of sludge by 10% to 20%.

Energy consumed in the feeding and mixing of conditioning chemicals is negligible in terms of overall wastewater-treatment-plant energy consumption. The energy required to produce the chemicals consumed (secondary energy requirement), however, may be significant and is summarized in Table 2-2.

Figures 2-1* through 2-12 present information on the costs of

*Illustrations for this chapter start on page 23.

various chemical-feeding systems for sludge conditioning based on the considerations discussed below. Table 2-3 shows the cost components.

Capital Costs

Polymer: Cost estimates of solution preparation and feeding equipment are based on the use of dry polymer. Chemical-feed equipment

TABLE 2-2
Energy Required to Produce Sludge-Conditioning Chemicals

	Fuel (10^6 BTU/ton)	Electricity (kwh/lb)
Ferric Chloride	10	0.5[a]
Lime (CaO)	5.5[a]	0.3
Polymer	3[a]	0.1

[a]Indicates principal type of energy used in production.

TABLE 2-3
Chemical Feed Systems Capital Cost (Sept. 1976)

Ferric chloride (pph)	52	520	5200
Feeder & storage equipment	$20,000	$ 64,000	$665,000
Housing	3,000	9,500	19,000
Electrical & instrumentation	4,750	15,000	166,000
Miscellaneous items	4,000	13,500	127,000
Total	$31,760	$102,000	$977,000
Polymer (pph)	0.4	3.5	35
Feeder & storage equipment	$ 7,800	$36,000	$273,500
Housing	3,000	6,400	25,500
Electrical & instrumentation	1,900	9,000	68,000
Miscellaneous items	1,900	7,700	55,000
Total	$14,600	$59,100	$422,000
Lime (pph)	52	520	5200
Feeder & storage equipment	$26,000	$125,000	$573,500
Housing	12,700	25,500	38,000
Electrical & instrumentation	6,500	31,000	143,000
Miscellaneous items	6,800	27,000	113,000
Total	$52,000	$208,500	$867,500

was based on feed of a 0.25% stock solution. Piping and buildings to house the feeding equipment and store the bags are included.

Systems with less than 0.4-pounds-per-hour capacity were based on the polymer's being manually fed to the mixing tank, with mixing and solution-feed equipment arranged for manual control. Two independent systems of tanks and feeders are included. Systems for 3.5 pounds per hour include two volumetric dry feeders discharging to two mixing tanks, arranged for batch control. The mixing tanks discharge to a single holding tank from which two solution feeders take their supply for application to the process. The system for 35 pounds per hour includes four volumetric dry feeders complete with steel day hoppers, dust collectors, bin gates, and flexible connectors. The system operation is automatic for the preparation and transfer of aged polymer solution from four mixing tanks to two holding tanks. Ten solution feeders meter the polymer to the treatment process.

Lime: Cost estimates were based on the use of hydrated lime for small plants (50 pounds per hour or less) and pebble quicklime for larger plants. Lime feed rates allow for peak rates of twice the average rate. Storage was provided for at least 15 days at the average rate. This storage time is arbitrary and will vary at each installation depending on the distance to and the reliability of the source of chemical supply. Piping and buildings to house the feeding equipment are included in the estimates. The estimated costs of steel bins with dust-collector vents and filling accessories are included.

Equipment for the 50-pounds-per-hour plant includes two volumetric feeder systems with manually loaded bins, dissolving chambers, and accessories. Equipment for the larger plants includes two gravimetric feeder-slaker systems. Steel bins are estimated to hold 1½ truckloads of quicklime each, and it is assumed that the truck will be equipped with a pneumatic blower. Equipment for a 5,200-pounds-per-hour rate includes four gravimetric feeder-slaker systems. The estimate includes the bin gate and feeder-slaker accessories and controls. Costs for the flow and pH controls include holding tanks with mixers, and slurry feeders with controls. Flow and pH-metering instruments are included.

Ferric Chloride: Costs of chemical storage and feeding equipment for rates of 36–3,600 pounds per hour have been estimated based on the use of liquid ferric chloride (4.75 pounds per gallon of $FeCl_3$).

Chemical feed equipment was sized for a peak feed rate of twice the average. Storage was provided for at least 15 days at the average feed rate. Piping and buildings to house the feeding equipment are included.

For the 36-pounds-per-hour rate, the feeding equipment for liquid ferric chloride includes two 18-gallons-per-hour hydraulic-diaphragm pumps (one operating and one standby) with the necessary accessories and equipment to pace the feed rate to the plant flow. Liquid ferric chloride is stored in two 3,000-gallon FRP tanks with accessories. The 6,000-gallon total capacity allows use of liquid ferric chloride in 4,000-gallon tank truck quantities.

For the larger rates, liquid ferric chloride is fed by 90-gallons-per-minute rotodip-type feeders (one standby) with the necessary accessories and control panel for proportioning chemical feed to flow.

Operating and Maintenance Costs

Ferric Chloride: The labor curve consists of two major components: (1) unloading of chemicals and (2) operation and maintenance (O & M) of chemical feeding equipment. Unloading requirements may be summarized as follows: 4,000-gallon trucks, 1.5 man-hours; 50,000-pound (dry bulk) trucks, 4 man-hours; 100,000-pound railcars, 6 man-hours; 50-pound bags (300 per truck), 8 man-hours; 50-gallon barrels (72 per truck), 9 man-hours. Metering pump O & M was estimated at 5 minutes per pump per shift. Dry feeder O & M was estimated at 10 minutes per feeder per shift.

Power requirements are based on use of a plunger-metering pump. Maintenance material costs are estimated at 3% of the equipment cost for all chemical-feed systems.

Polymer: Unloading requirements were based on 8 man-hours per 300 50-pound bags, and O & M requirements on 385 man-hours per year per feeder. Mixing labor was estimated at 10 hours per 1,000 pounds of polymer up to 10 million gallons per day and 3 hours per 1,000 pounds over 10 million gallons per day.

Power requirements for polymer were based on use of plunger-metering pumps and 6.4 horsepower-hours for mixing 100 pounds of polymer.

Lime: Lime unloading requirements are included in the labor curve. The other major labor components are related to O & M of the slaking and feeding equipment. These components were estimated as follows: slaker, 1 hour per shift per slaker in use; feeder, 10 minutes per hour per feeder; slurry pot, feed line (for slaked lime), 4 hours per week.

Power requirements for lime feed were determined by analyses of several plants. The major components and the values used in the curves, all expressed as kilowatt-hours per 1,000 pounds of lime feed: slakers, 1.6–0.8; bin activators, 2.7–0.36; grit conveyors, 0.45–0.06; dust collection fans, 0.04–0.02; slurry mixers, 0.027–0.020; slurry feed pumps, 2.2–1.4.

HEAT TREATMENT

There are two basic types of high-temperature/high-pressure treatment of sludges. One, "wet-air oxidation," involves the flameless oxidation of sludges at 450°–550°F at pressures of about 1200 psig. The other type of heat treatment is carried out at lower temperatures and pressures (350°–400°F at 150–300 psig) to improve the dewaterability of sludges, and is the subject of this section.

When colloidal gel systems are heated, thermal activity causes water to escape from the structure. It is the goal of heat-treatment systems to release bound water from the sewage sludge to improve the dewatering and thickening characteristics of the sludge. Unfortunately, the physical effect of heat treatment also ruptures the cell walls of biological sludges, releasing bound organic colloidal material, solubilizes previously insoluble organic material, and creates fine particulate debris. This solubilization process means that a principal result of heat treatment is the conversion of suspended solids to dissolved or dispersed solids, facilitating dewatering but simultaneously creating a separate problem of recycling highly polluted liquid from the dewatering process to the wastewater-treatment plant. This recycling must be recognized when assessing the feasibility of heat treatment.

Heat treatment is a conditioning technique which has had increasing use by consulting engineers, although its use is still limited when compared to the large number of plants using other techniques for sludge conditioning. Several operating heat-treatment plants have had significant operational problems from (1) the increased loadings of biochemical oxygen demand (BOD), chemical oxygen demand (COD), nitrogen, phosphorus, and suspended solids on the secondary plant; (2) odors from off-gases from the process; and (3) the refractory nature of a portion of the recycled load which will pass through the secondary plant as COD and can cause taste and odor problems in downstream water plants. The COD refractory to the biological processes has also been found difficult to remove by advanced wastewater-treatment processes such as activated-carbon adsorption. The refractory nature of the recycled organics may limit the applicability of this conditioning process in areas where downstream water uses dictate very low effluent COD values.

An advantage of the heat treatment of sludges is that it produces a more readily dewaterable sludge than chemical conditioning. Dewatered sludge solids of 30%–40% (as opposed to 15%–20% with chemical conditioning) have been achieved with heat treatment at relatively high loading rates on the dewatering equipment (two to three times the rates with chemical conditioning). The process also provides effective disinfection of the sludge.

A typical heat treatment process (Zimpro LPO) is shown in Figure 2-13. Sludge is ground and pumped to a pressure of about 300 psi. Compressed air is introduced into the sludge and the mixture is brought to an operating temperature of about 350°F by heat exchange and direct steam injection, and flows to the reactor. The heated, conditioned sludge is cooled by heat exchange with the incoming sludge. The treated sludge is separated by settling before the dewatering step. Gases released at the separation step are passed through a catalytic afterburner at 650°–750°F. The system is more mechanically complex than many unit processes in municipal wastewater-treatment plants, and some installations have encountered significant maintenance problems.

In order to estimate the costs to construct thermal-treatment facilities of various capacities, construction cost data were obtained from the records of approximately 30 thermal-treatment plants, from manufacturers, and from engineering estimates. Bidding information and records from plants frequently did not contain breakdowns indicating what portions of the cost were for thermal treatment, and where some breakdown was provided it often covered the total costs for sludge handling, including prethickening, storage, thermal treatment, decanting, vacuum-filter or centrifuge dewatering, incineration, and engineering. Information from some plants included the costs for buildings, and at times these buildings were quite elaborate and large, including room for expansion and housing dewatering and incineration systems as well as thermal-treatment facilities. Building costs for some plants were reported to be over two-thirds of the costs for thermal-treatment equipment.

Because of these various differences in what was included in reported costs, the costs for thermal treatment alone were separated for comparison and plotting. The resulting costs for thermal treatment include sludge feed pumps, grinders, heat exchangers, reactors, boilers, gas separators, air compressors (where applicable), decanting tanks, standard odor-control systems, and piping, controls, wiring, and installation services usually furnished by the equipment or system manufacturer. Not included in the basic thermal-treatment costs are buildings; footings; piping, electrical work, and utilities not supplied by the equipment manufacturer; sludge storage and thickening prior to thermal treatment; sludge dewatering, incineration, or disposal; land; and engineering fees. Costs are plotted as cost per unit of thermal treatment capacity versus thermal treatment capacity (dollars per gallon per minute vs. gallons per minute) in Figure 2-14 and as cost for thermal treatment versus thermal treatment capacity (dollars vs. gallons per minute) in Figure 2-15.

The curves show a rapid rise in unit construction costs for plants smaller than about 20 gallons per minute, indicating that there is a limiting plant size below which high cost makes the process infeasible.

Typical fuel requirements (shown in Figure 2-16) averaged 900-1000 BTUs per gallon for plants not practicing air addition and 300-600 BTUs per gallon, depending on the degree of oxidation obtained, for plants practicing air addition. Figure 2-17 presents annual electrical-energy requirements which were found to be about 22 kilowatt-hours per 1,000 gallons for plants practicing air addition and 10 kilowatt-hours per 1,000 gallons for plants without air addition. Labor requirements are presented in Figure 2-18.

Figure 2-19 presents annual costs for materials and supplies. Curve A on Figure 2-19 shows the normal annual cost for materials and supplies required to operate and maintain the thermal-treatment system. These costs are plotted against thermal-treatment-plant capacity and include materials and parts such as seals, packing, coatings, lamps, bearings, grinder blades, and other items used in scheduled and normal maintenance. They also include operating supplies such as lubricants, cleaning chemicals, boiler feedwater, and water-treating chemicals. Besides the normal periodic maintenance required for a plant and covered by curve A, additional costs for major overhaul work are incurred. This work includes such items as motor rewinding; major overhauls of pumps and compressors; major nonroutine rehabilitation or replacement of heat exchanger tubing, piping, and controls; and refitting boilers. This type of work is required at an average interval of about six to seven years, depending on the variety of conditions at a particular plant. Because labor for this type of major work is often contracted, labor costs are treated as part of the overhaul and are included in its cost. Curve B shows the combination of these costs with those included under curve *A* to give the total annual cost for materials and supplies.

A very important consideration in evaluating the total costs of heat treatment is the costs associated with handling and treating the strong liquors resulting from thermal processing of sludge and with treating the odorous off-gases from the processes. The methods most commonly used and most generally effective for controlling odors from thermal treatment are high-temperature incineration, adsorption on activated carbon, and chemical scrubbing. The costs shown for these methods in Figures 2–20 and 2–21 include costs for collection of gases; ducting; fans; chemical feeding, mixing, and storage equipment; automatic control systems; disposal of removed and waste materials; and discharge of treated gases as well as odor removal itself. The incineration or afterburning process considered consists of pretreatment by water scrubbing using treated effluent in a packed bed and direct-flame incineration at 1,500°F with recovery of 40% of the input heat. The carbon adsorption process includes prescrubbing with effluent, dual-bed adsorption or activated carbon, regeneration of carbon with low-pressure steam, condensation of vapors, and incineration of the waste organic stream. The chemical-scrubbing system utilizes three stages of scrubbing in packed beds. The first two stages use secondary effluent and a final stage uses a buffered potassium permanganate solution.

From the standpoint of total cost for odor control, incineration and scrubbing are the least expensive methods for use in very small plants. As size increases, however, total costs for scrubbing and adsorption quickly become and remain almost equal, and end up approximately half the total cost for incineration. It must be emphasized that odor-control systems for use in thermal-treatment plants should be selected on the basis of what will do the job. Only after it is determined that more than one process will perform adequately can selection be made on the basis of cost.

An estimate of the total costs of thermal-treatment systems—including both direct and indirect costs—for odor control and incremental treatment costs for handling recycle liquors is shown in Table 2–4.

In evaluating costs, the costs of the total system (conditioning, dewatering, and disposal) must be considered. The fact that heat treatment produces a more readily dewaterable sludge produces economies in downstream dewatering processes. It may produce a sludge dry enough to burn without auxiliary fuel, providing downstream savings.

TABLE 2-4
Summary of Direct and Indirect Cost for Thermal Treatment (March 1975)

Sludge[a] (ton/day)	Construction Cost[b]			O & M Cost			Total Cost
	Direct	Indirect	Total	Direct	Indirect	Total	
1	97.53	4.11	101.64	150.14	4.93	155.07	256.71
5	30.79	3.18	33.97	46.46	3.67	50.13	84.10
10	21.45	2.93	24.38	32.52	3.50	36.02	60.40
50	12.20	1.83	14.03	19.10	2.99	22.09	36.12
100	10.96	1.98	12.94	16.58	2.87	19.45	32.39

[a]Basis: (a) 1.1 tons solids per mdg (includes recycle); (b) 4 gpm to thermal treatment per mdg @ 4½% solids; (c) 8,000 hr/yr operating time.
[b]Amortized over 20 years @ 7%; all costs in dollars per ton; does not include odor control.

OTHER TECHNIQUES

Freezing

A number of people have observed that sludge frozen and later thawed in sand drying beds or lagoons had good dewatering and fertilizer or soil-conditioning characteristics. Thawed sludge was stable and dewatered rapidly if provisions were made for water drainage.[14] These observations encouraged researchers, particularly in Great Britain, to evaluate artificially freezing sludge as a means of promoting rapid dewatering.

The city of Milwaukee, Wisconsin, has studied the process for application to activated sludge.[15,16] These studies found that vacuum filter rates of 55 pounds per square foot per hour were achievable; that the filtrate and filter cake quality were equivalent or better than that produced from a conventional vacuum-filter operation; that freeze-conditioned sludge could be dewatered by a wire screen cloth (40-80 mesh) by gravity draining, and that freezing rate was an important variable.

Early engineering studies at Milwaukee revealed that equipment costs and space requirements would be substantially higher than for chemical-conditioning techniques currently employed, and that the space requirement for freeze conditioning would be 65-130 times that required for the conventional chemical-dewatering system. Equipment costs for the freeze-conditioning system were estimated at 7–10 times those for the chemical-conditioning process. The estimated annual operating costs for the freeze-conditioning process were three to four times that of the chemical conditioning approach.

Coupled with the high capital cost, its very high operating cost has been the major reason why freezing has not been adopted as a conditioning technique for wastewater sludge.

Hydrolysis with Sulfur Dioxide

This process consists of heating activated sludge in the presence of water and small amounts of sulfur dioxide to improve its dewatering characteristics. This treatment increases soluble solids and produces a filtrate which can be concentrated to produce a molasses-type syrup which could be of value as an animal feed. [17-19]

A small-scale study has been made to develop a preliminary evaluation of the feasibility of the hydrolysis process applied to sewage sludge. It was found that the filtration rate of activated sludge was increased by a factor of six when SO_2 was added to the activated sludge before heat treatment while heat treatment alone increased the rate by a factor of only three. The amount of moisture retained in the filter cake was also reduced by the SO_2 treatment. It was found that heat treatment alone increases the soluble-solids content of activated sludge by about 90%, and that an additional 20% was obtained by the addition of SO_2 (0.5% sulfurous acid, 140°C for 1 hour). Evaporation of the filtrate to a syrup with 60% solids produced a molasses which was 82% organic.

Preliminary cost estimates indicated that the sale of the molasses resulting from this process could recover about 20% of the cost of the hydrolysis treatment. The SO_2 hydrolysis process has not yet been tried on sewage sludges on a plant scale or continuous basis. Thus, the economics of the process and the marketability of the resulting molasses are yet to be demonstrated.

Radiation Treatment

Radiation of sludges produces charged and oxidizing species which affect colloidal systems and may improve the thickening and dewatering characteristics of the sludges.[20] A limited amount of data are available on the feasibility of using radiation treatment for sludge conditioning. [20-23] A preliminary analysis of the economic feasibility indicates that if a dosage of 10^5 rads would enable a doubling of the vacuum-filtration rate, then the costs of radiation treatment would be of the same order of magnitude as the potential savings.

A study of the effects of gamma irradiation on the settleability and filterability of digested activated sludge indicated that a dose of 5×10^5 rads showed essentially no effect on the settling properties of the sludge, but in combination with ferric chloride conditioning to approximately one-third of the optimum conditioning dosage, it was able to effect about a threefold increase in the dewaterability. The filterability of undigested activated sludge was not increased by the same treatment.

Another study found that irradiation produced a marked effect on the filterability of sludge, but that this effect saturated at a dose of approximately 10^5 rads at a level of specific resistance too high to permit the sludges to be filtered at a useful rate on a rotary vacuum filter.[22] The range of specific resistance needed for effective vacuum filtration could not be reached with the use of ionizing radiation alone.

The use of radiation for sludge conditioning is not currently practiced in the U.S. and it does not appear likely that it will be in the near future.

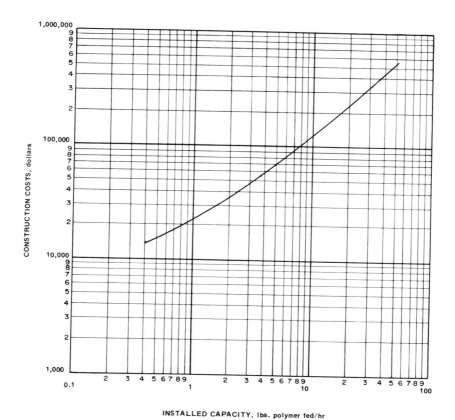

Figure 2-1. Construction costs for polymer storage and feeding (Sept. 1976).

24/Handbook of Sludge-Handling Processes

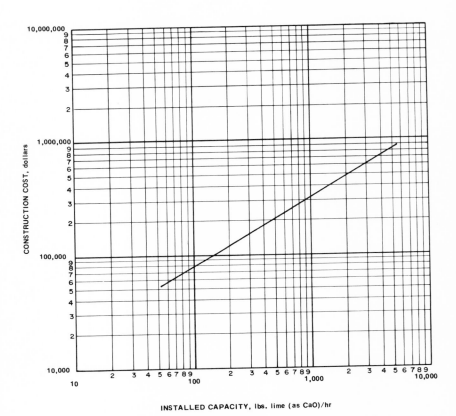

Figure 2–2. Construction costs for lime storage and feeding (Sept. 1976).

Sludge Conditioning/25

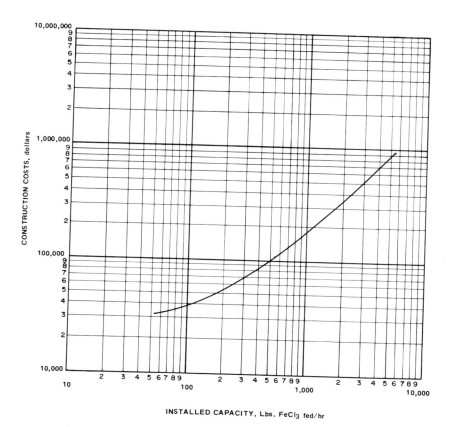

Figure 2-3. Construction costs for ferric chloride storage and feeding (Sept. 1976).

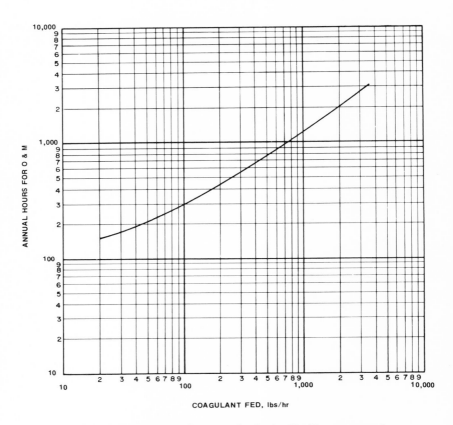

Figure 2-4. Labor requirements for ferric chloride storage and feeding.

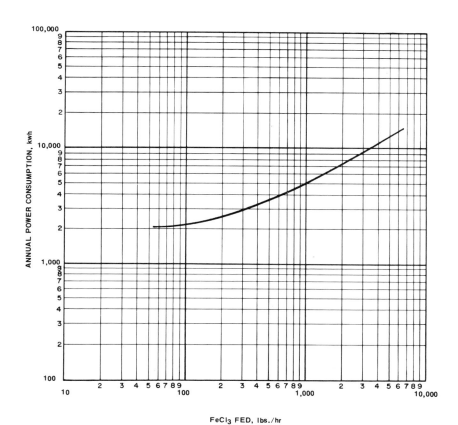

Figure 2-5. Power requirements for ferric chloride feeding.

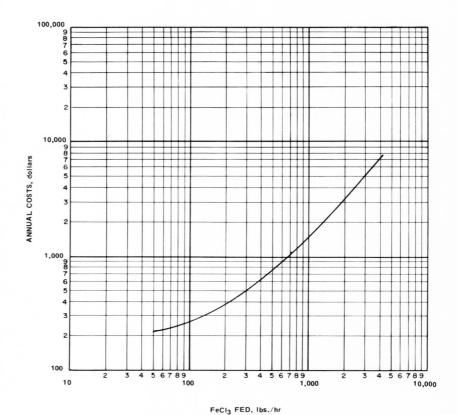

Figure 2-6. Maintenance material costs for ferric chloride storage and feeding (Sept. 1976).

Sludge Conditioning/29

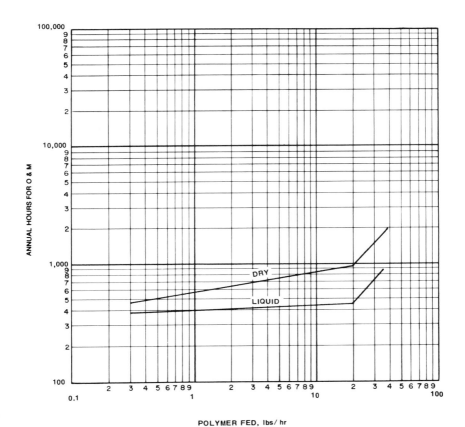

Figure 2-7. Labor requirements for polymer feeding.

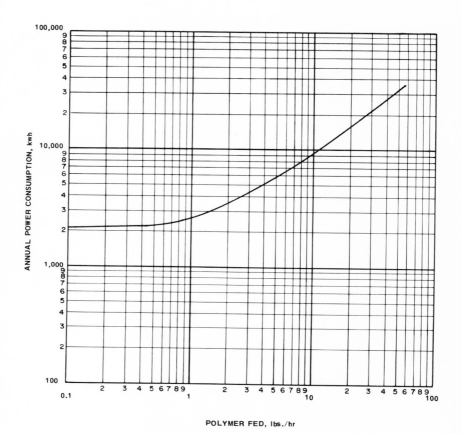

Figure 2-8. Power requirements for polymer mixing and feeding.

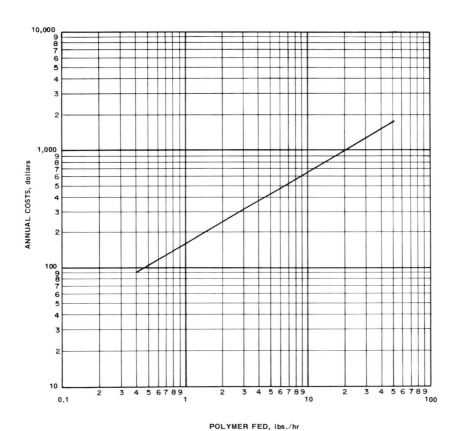

Figure 2-9. Maintenance material costs for polymer storage and feeding (Sept. 1976).

32/Handbook of Sludge-Handling Processes

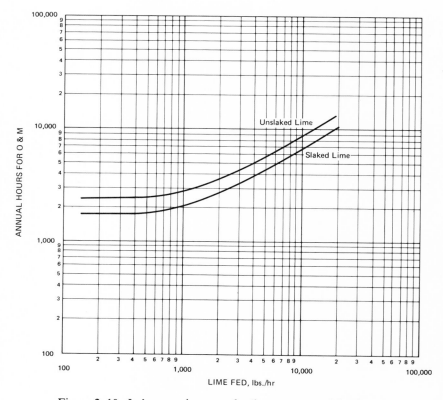

Figure 2-10. Labor requirements for lime storage and feeding.

Sludge Conditioning/33

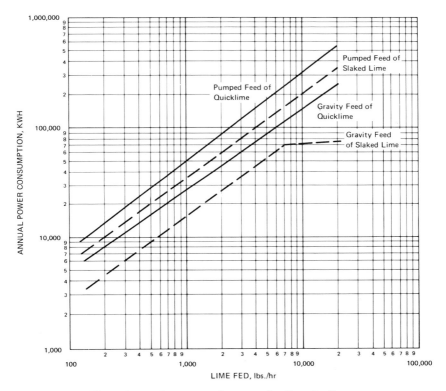

Figure 2-11. Power requirements for lime feeding.

34 / Handbook of Sludge-Handling Processes

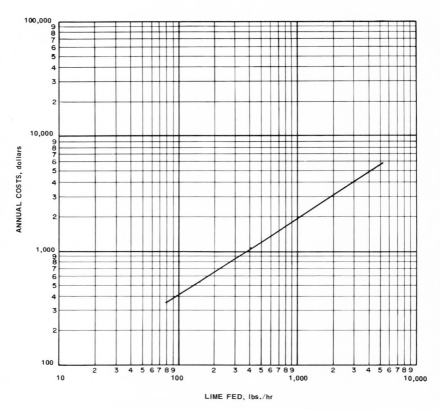

Figure 2-12. Maintenance material costs for lime storage and feeding (Sept. 1976).

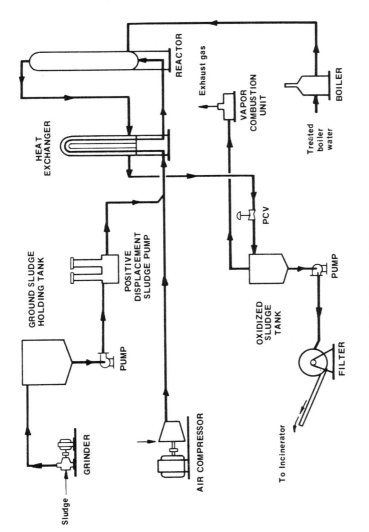

Figure 2-13. Zimpro LPO system.

36 / Handbook of Sludge-Handling Processes

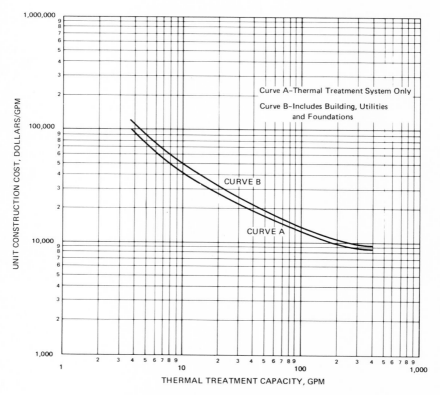

Figure 2-14. Direct construction costs for thermal treatment per unit of capacity (March 1975).

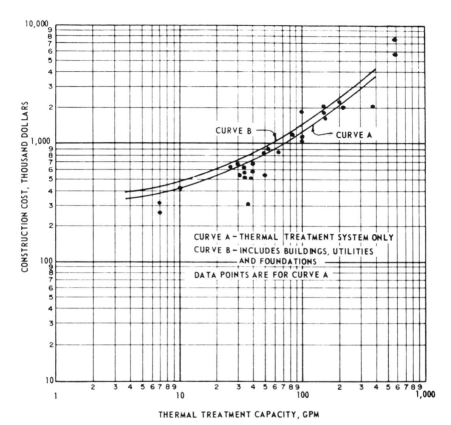

Figure 2–15. Direct construction costs for thermal treatment (March 1975).

38 / Handbook of Sludge-Handling Processes

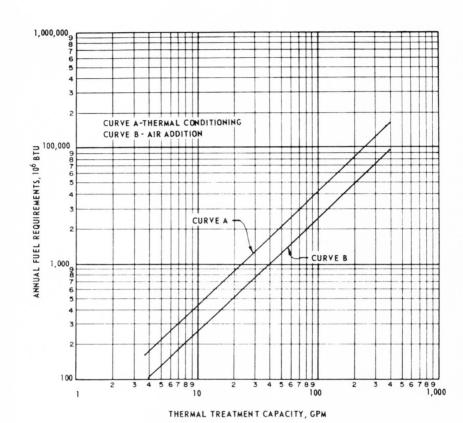

Figure 2–16. Annual direct fuel requirements for thermal treatment.

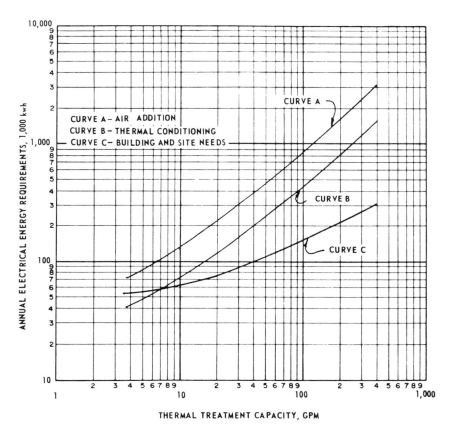

Figure 2-17. Annual direct electrical energy requirements for thermal treatment.

40/Handbook of Sludge-Handling Processes

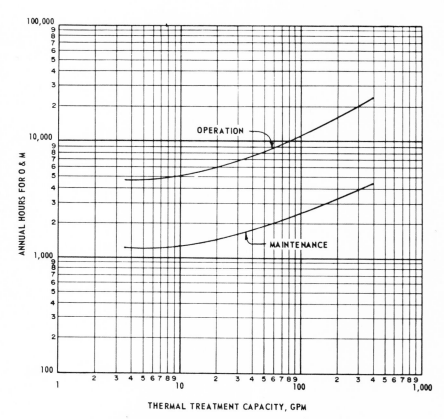

Figure 2-18. Labor requirements for thermal treatment.

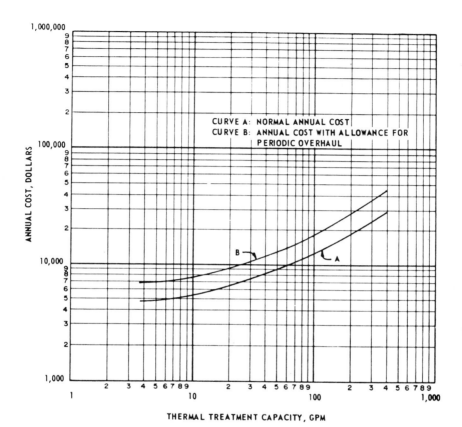

Figure 2-19. Materials and supplies costs for thermal treatment.

42/Handbook of Sludge-Handling Processes

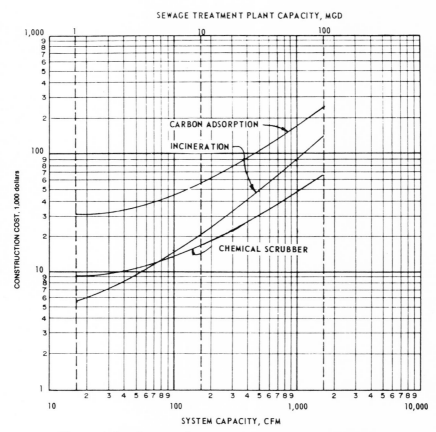

Figure 2-20. Construction cost for odor control systems (March 1975).

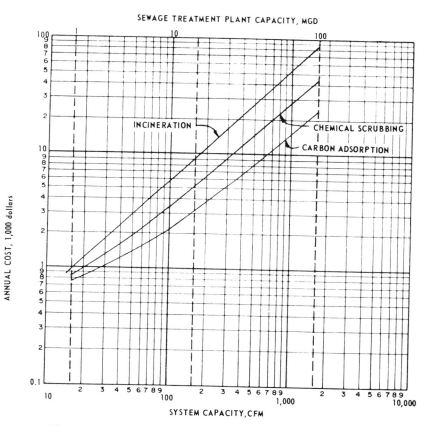

Figure 2–21. Operation and maintenance costs for odor control systems (March 1975).

Chapter 3

Sludge Thickening, Pumping and Storage

The purpose of sludge thickening is to reduce the sludge volume to be stabilized, dewatered, or disposed of. Figure 3-1* illustrates the impact that thickening can have on sludge volume. Thickening a 1% sludge to 6% solids reduces the volume of sludge to be handled by a factor of over five. This reduction can provide significant savings in the cost of dewatering, digestion, or other downstream facilites. There are three commonly used methods for sludge thickening: gravity, flotation, and centrifugation.

GRAVITY THICKENING

Thickening by gravity is the most common concentration process in use at wastewater-treatment plants. It is simple and inexpensive. Gravity thickening is essentially a sedimentation process similar to that which occurs in all settling tanks. But in comparison with the initial waste-clarification stage, the thickening action is relatively slow. The theoretical aspects of gravity thickening have been the subject of many studies and are well summarized in a few recent papers.[24-28]

Figure 3-2 illustrates a typical circular gravity thickener. The units have a typical side water depth of 10 feet. Loading rates are expressed in terms of pounds of dry solids in the sludge applied to the thickener surface area per day (pounds per day per square foot). Table 3-1 summarizes typical results achieved with gravity thickening.

The degree to which waste sludges can be thickened depends on many factors, among the most important, the type of sludge being thickened and its volatile-solids concentration. Bulky biological sludge, particularly from the activated-sludge process, will not concentrate to the same degree as raw primary sludge. Activated sludges, if thickened

*Illustrations for this chapter start on page 52.

TABLE 3-1
Typical Results of Gravity Thickening

Sludge type	Feed solids concentration (%)	Typical loading rate (lb/sq ft/day)	Thickened sludge concentration (%)
Primary	5.0	20–30	8.0–10
Trickling filter	1.0	8–10	7–9
Primary + FeCl$_3$	2.0	6	4.0
Primary + low lime	5.0	20	7.0
Primary + high lime	7.5	25	12.0
Primary + WAS[a]	2.0	6–10	4.0
WAS[a]	1.0	5–6	2–3
Primary + (WAS[a] + FeCl$_3$)	1.5	6	3.0
(Primary + FeCl$_3$) + WAS[a]	1.8	6	3.6
Digested primary	8.0	25	12.0
Digested primary + WAS[a]	4.0	15	8.0
Digested primary + (WAS[a] + FeCl$_3$)	4.0	15	6.0
Tertiary, two-stage high lime	4.5	60	15.0
Tertiary, low lime	3.0	60	12.0

[a] WAS = waste activated sludge.

separately, are usually thickened by the flotation process. The degree of biological treatment and the ratio of primary to secondary (biological) sludge will affect the ultimate solids concentration obtained by gravity thickening. Hydraulic and surface loading rates are also of importance. Current practice calls for the use of overflow rates of 400–800 gallons per day per square foot. Excessively low flow rates can lead to odor problems. If the sludge flow to the thickener is far below the design rate, pumping secondary effluent to the thickener may be practiced to minimize odors.

The quality of the overhead liquid removed from the sludge solids is important in any thickening operation because this liquid is usually returned to the treatment processes. Generally, the overhead quality is similar to that of raw sewage, 150–300 milligrams per liter of suspended solids and a BOD of about 200 milligrams per liter. A well-operated thickener should have a minimum of anaerobic decomposition and a solids capture exceeding 90%. Thus, the overflow returned to the treatment process should not present an operational problem.

Figures 3–3 through 3–6 summarize the cost of gravity thickeners. Table 3–2 presents the capital cost components. The costs are typically $2–$5 per ton of dry solids.

TABLE 3-2
Estimated Construction Costs of Gravity Type Sludge Thickeners (Sept. 1976)

Cost component	Surface area (sq ft)				
	315	710	1,590	2,830	3,850
Manufactured equipment	$33,500	$ 44,200	$ 54,300	$ 58,500	$ 62,800
Concrete	2,200	3,500	5,700	8,100	10,000
Steel	8,500	13,200	20,300	27,700	32,400
Labor	6,700	10,800	16,300	22,200	27,000
Electrical and instrumentation	7,600	10,800	14,500	17,500	19,800
Miscellaneous items	8,800	12,400	16,700	20,100	22,800
Contingency	10,100	14,200	19,200	23,100	26,200
Total Estimated Cost	$77,400	$109,100	$147,000	$177,200	$201,000

FLOTATION THICKENING

Flotation thickening units are becoming increasingly popular for sewage treatment plants in the U.S., especially for handling waste-activated sludges (WAS) since they have the advantage over gravity-thickening tanks of offering higher solids concentrations and lower initial cost for the equipment. The objective of flotation thickening is to attach a minute air bubble to suspended solids and cause the solids to separate from the water in an upward direction because the solid particles have a specific gravity lower than that of water when the bubble is attached.

Figure 3-7 illustrates the basic considerations involved in the process. A portion of the unit effluent, or plant effluent, is pumped to a retention tank (a pressurization tank) at 60–70 psig. Air is fed into the pump discharge line at a controlled rate and mixed by the action of an eductor driven by the reaeration pump. The flow through the recycle system is metered and controlled by a valve located immediately before the mixing with the sludge feed. The recycle flow and sludge feed are mixed in a chamber at the unit inlet. If flotation aids (such as polymers) are employed, they are normally introduced in this mixing chamber. The sludge particles are floated to the sludge blanket and the clarified effluent is discharged under a baffle and over an adjustable weir. The thickened sludge is removed by a variable-speed skimming mechanism. In practice, bottom sludge collectors are also furnished for removal of any settled sludge or grit that may accumulate. Sludge thickening occurs in the sludge blanket, which is normally 8–24 inches thick. The buoyant sludge and air bubbles force the surface of the blanket above the water level, inducing drainage of water from the sludge particles.

Similarly to gravity thickening, the type of quality of sludge to be floated affects the unit performance. Flotation thickening is, as stated before, most applicable to activated sludges, but higher float concentrations can be achieved by combining primary with activated sludge. Equal or greater concentrations may be achieved by combining sludges in gravity thickening units, however. A high Sludge Volume Index (SVI), representing a bulky sludge, results in poor thickener performance. Table 3–3 presents typical results from flotation thickening.

Many different chemicals have been used in various air-flotation systems. The overall effect is to increase the allowable solids loadings, increase the percentage of floated solids, and increase the clarity of the effluent. Cationic polyelectrolytes have been the most successful chemical used in sewage sludge thickening,[29–32] with dosages of 8–12 pounds per ton reported as typical. Waste activated sludge can typically be thickened to 3%–5% solids without chemicals and 4%–6% with chemicals.

TABLE 3–3
Flotation Thickener Operation and Performance

Operation parameter	Range	Typical
Solids loading (lb dry solids/hr/sq ft of surface)		
With chemicals	2 to 5	2
Without chemicals	1 to 2	1
Influent solids concentration (mg/l)	5,000 min	5,000 min
Air-to-solids ratio	0.02–0.04	0.03
Blanket thickness (inches)	8–24	—
Retention tank pressure (psi)	60–70	—
Recycle ratio (% of influent flow)	30/150	
Expected performance		
Float solids concentration (%)		3–7
Solids removal (%)		
With flotation aid		95
Without flotation aid		50–80

Figures 3-8 through 3-11 present cost information on flotation thickening with capital cost components shown in Table 3-4. Costs are typically $7–$15 per ton of dry solids.

CENTRIFUGAL THICKENING

Although centrifuges have been used widely for dewatering (see chapter 4), they have had limited use for thickening because of their relatively high cost. They have been used for thickening of WAS where space limitations or sludge characteristics make other methods unsuitable.[6] WAS concentrations of 5%–8% have typically been produced by centrifugal thickening.

SLUDGE PUMPING

Figures 3-12 through 3-15 and Table 3-5 summarize costs for sludge-pumping stations using positive-displacement pumps. The station is based on an underground structure housing pumping units and piping, constructed adjacent to and in conjunction with the solids-separation-

TABLE 3-4
Estimated Construction Costs of Flotation Type Sludge Thickeners (Sept. 1976)

Cost component	Surface area (sq ft)			
	400	800	1,600	3,200
Manufactured equipment	$100,000	$120,000	$152,400	$298,000
Concrete	6,400	8,500	12,900	21,800
Steel	17,800	26,100	37,100	59,200
Labor	42,400	55,300	76,400	132,400
Metal pipe and valves	2,100	4,200	5,700	6,500
Electrical and instrumentation	33,700	42,800	56,900	103,600
Miscellaneous items	30,400	38,500	51,200	93,200
Contingency	34,900	44,300	58,900	107,200
Total Estimated Cost	$267,700	$339,700	$451,500	$821,900

TABLE 3–5
Estimated Construction Costs of Sludge Pumping Stations (Sept. 1976)

Cost component	Firm pumping capacity (gpm)						
	50	100	200	500	1,000	2,000	5,000
Manufactured equipment	$ 7,200	$10,300	$ 15,200	$ 23,400	$ 32,800	$ 46,800	$ 72,600
Concrete	1,300	1,900	2,800	4,400	6,100	8,700	13,500
Steel	4,000	5,800	8,600	13,200	18,500	26,500	41,000
Labor	13,100	19,000	27,900	42,900	60,100	85,900	133,100
Metal pipe and valves	7,700	11,000	16,300	25,100	35,100	50,200	77,800
Housing	8,400	12,200	18,000	27,700	38,700	55,300	85,800
Electrical and instrumentation	8,300	12,000	17,700	27,300	38,200	54,500	84,500
Miscellaneous items	7,500	10,800	15,900	24,500	34,300	48,900	75,800
Contingency	8,600	12,400	18,400	28,300	39,600	56,500	87,600
Total Estimated Cost	$66,100	$95,400	$140,800	$216,800	$303,400	$433,300	$671,700

TABLE 3-6
Estimated Construction Costs of Sludge Holding Tanks (Sept. 1976)

Cost component	Volume (1,000 cu ft)			
	3	15	50	200
Manufactured equipment	$ 3,300	$ 20,000	$ 40,000	$150,000
Concrete	2,600	6,600	12,200	36,200
Steel	4,700	7,400	13,700	40,600
Labor	13,000	32,000	63,000	156,500
Metal pipe and valves	1,300	3,000	3,000	10,000
Electrical and instrumentation	3,700	10,400	19,800	59,000
Miscellaneous items	4,300	11,900	22,800	67,800
Contingency	4,900	13,700	26,200	78,000
Total Estimated Cost	$37,800	$105,000	$200,700	$598,100

unit process. A superstructure is included to access the station from the ground level and to house electrical control equipment.

SLUDGE HOLDING TANKS

Sludge holding tanks are a necessity where sludge is to be dewatered. The use of holding tanks serves to dampen fluctuation in the quantity and quality of sludge flow to the dewatering process and blends sludge from various sources. Holding tanks may also be used to concentrate sludge by decanting liquids after stratification of sludge in the tank.

Construction costs are closely related to the volume of the tank. Costs shown in Figure 3-16 and Table 3-6 include mixers for the tank, but do not include sludge pumping from the tank. Figures 3-16 through 3-19 summarize costs for sludge holding tanks.

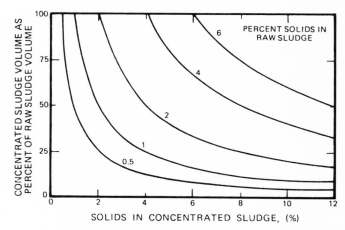

Figure 3-1. Effect of increasing sludge solids on the final sludge volume (reference 6).

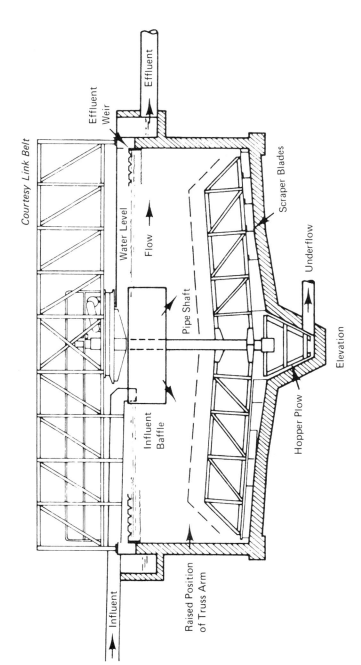

Figure 3-2. Gravity thickener (reference 6).

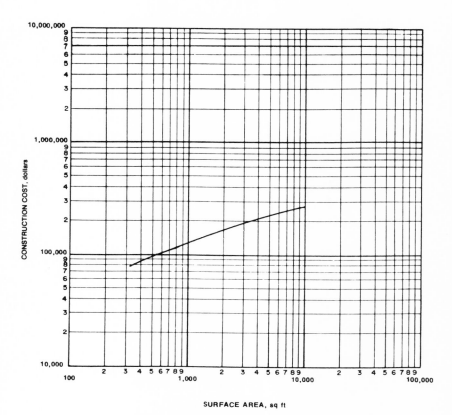

Figure 3-3. Construction costs for gravity sludge thickeners (Sept. 1976).

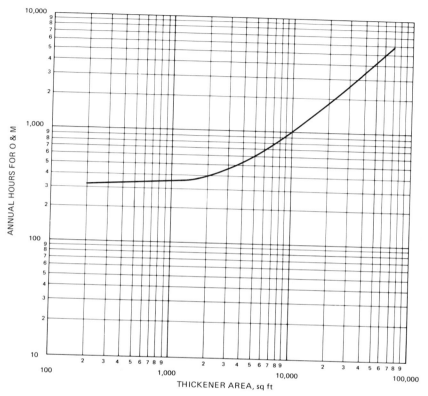

Figure 3-4. Labor requirements for gravity thickening.

56 / Handbook of Sludge-Handling Processes

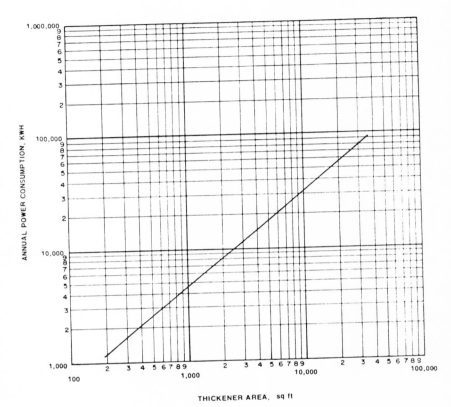

Figure 3-5. Power requirements for gravity thickening.

Sludge Thickening, Pumping, and Storage/57

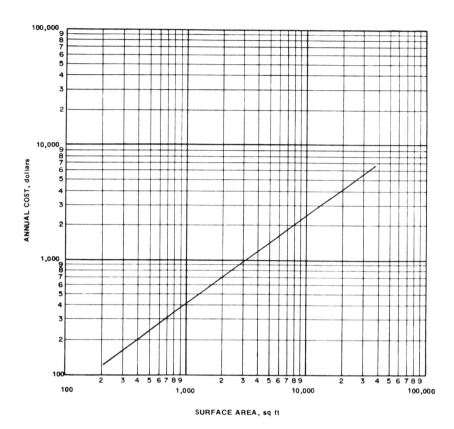

Figure 3-6. Maintenance materials and supply costs for gravity sludge thickeners (Sept. 1976).

58/Handbook of Sludge-Handling Processes

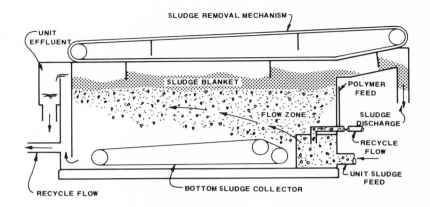

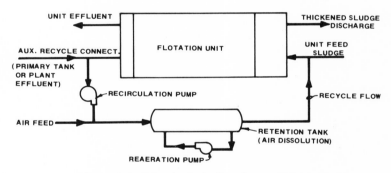

Figure 3–7. Dissolved air flotation system.

Sludge Thickening, Pumping, and Storage/59

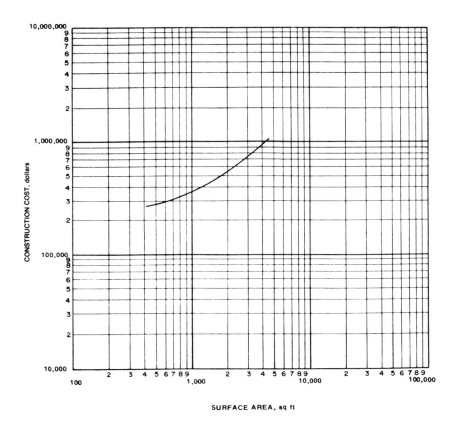

Figure 3-8. Construction costs for flotation sludge thickeners (Sept. 1976).

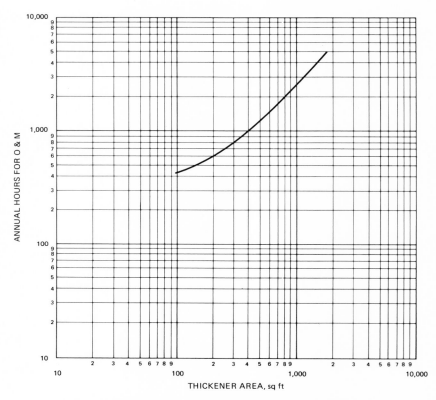

Figure 3-9. Labor requirements for flotation thickening.

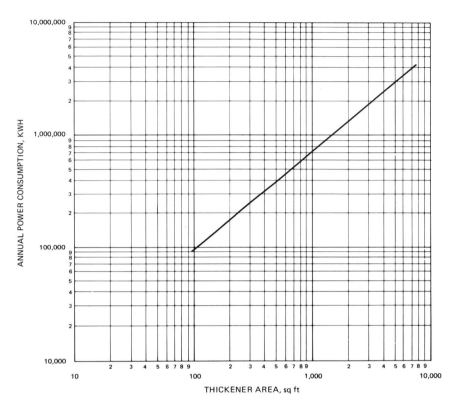

Figure 3-10. Power requirements for flotation thickening.

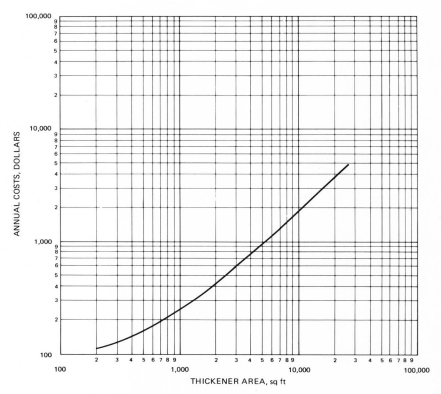

Figure 3-11. Maintenance material costs for flotation thickening (Sept. 1976).

Sludge Thickening, Pumping, and Storage/63

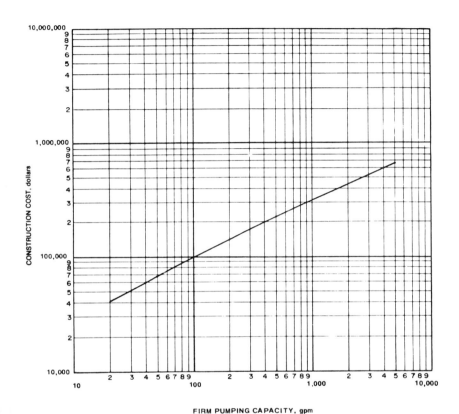

Figure 3-12. Construction costs for sludge pumping stations (Sept. 1976).

64/Handbook of Sludge-Handling Processes

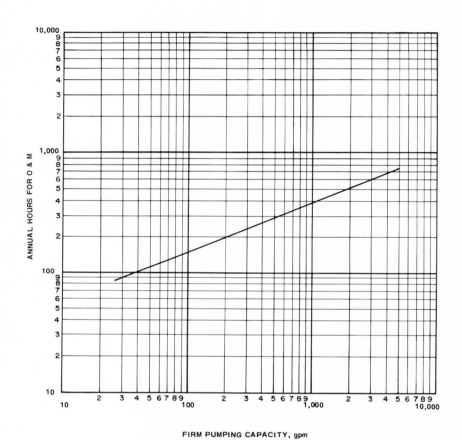

Figure 3-13. Labor requirements for sludge pumping.

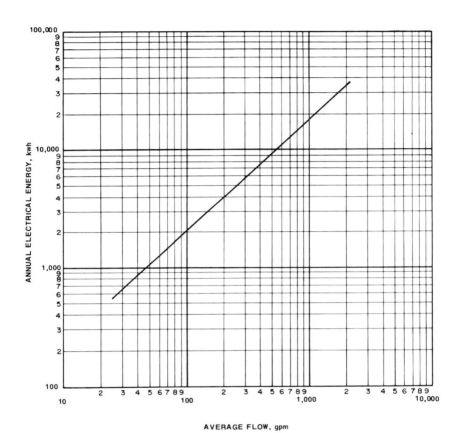

Figure 3-14. Energy requirements for sludge pumping.

66/Handbook of Sludge-Handling Processes

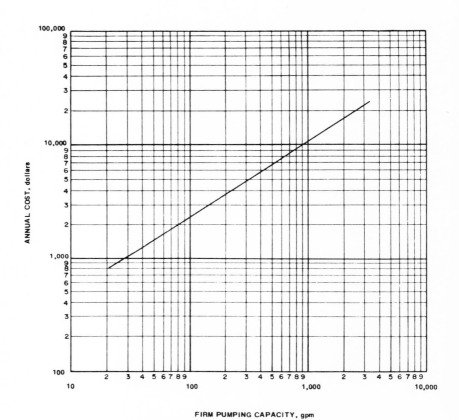

Figure 3-15. Maintenance material and supply costs for sludge pumping (Sept. 1976).

Sludge Thickening, Pumping, and Storage/67

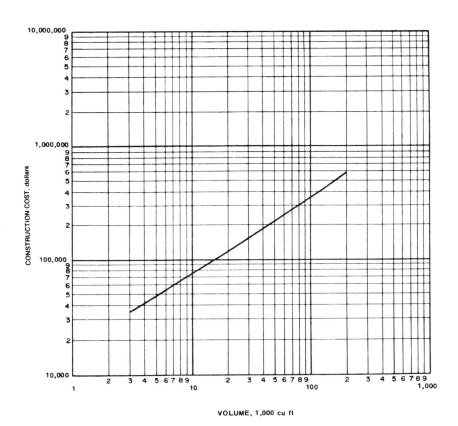

Figure 3-16. Construction costs for sludge holding tanks (Sept 1976).

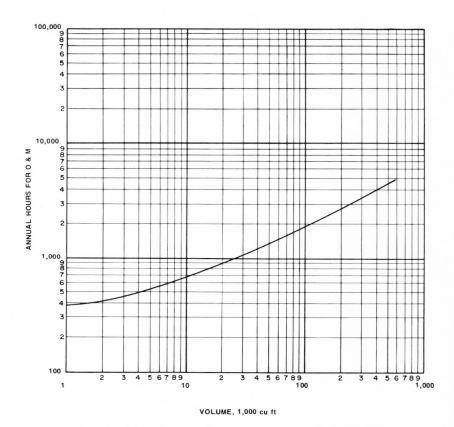

Figure 3-17. Labor requirements for sludge holding tanks.

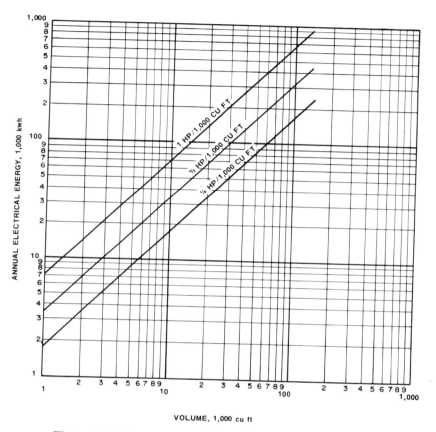

Figure 3-18. Energy requirements for sludge holding tanks.

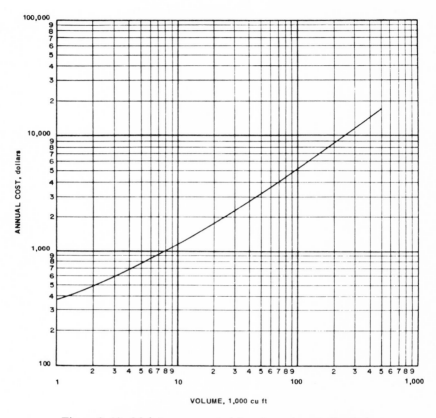

Figure 3-19. Maintenance material and supply costs for sludge holding tanks (Sept. 1976).

Chapter 4

Sludge Dewatering

DRYING BEDS

The most widely used dewatering method in the United States is drying the sludge on open or covered sandbeds. Over 6,000 wastewater-treatment plants use this method.[6] They are especially popular in small plants. Sandbeds possess the advantage of needing little operator skill. Air drying is normally restricted to well-digested sludge, because raw sludge is odorous, attracts insects, and does not dry satisfactorily when applied at reasonable depths. Oil and grease discharged with raw sludge clog sandbed pores and thereby seriously retard drainage. The design and use of drying beds are affected by many parameters. These include weather conditions, sludge characteristics, land values and proximity of residences, and use of sludge-conditioning aids. Climatic conditions are most important. Factors such as the amount and rate of precipitation, percentage of sunshine, air temperature, relative humidity, and wind velocity determine the effectiveness of air drying. It is important that wastewater sludge be well digested for optimum drying. In well-digested sludge, entrained gases tend to float the sludge solids and leave a layer of relatively clear liquid, which can readily drain through the sand. Typical design criteria are found in Table 4-1.

Sandbeds can be enclosed by glass. Glass enclosures protect the drying sludge from rain, control odors and insects, reduce the drying periods during cold weather, and can improve the appearance of a waste-treatment plant. Experience has shown that only 67%–75% of the area required for an open bed is needed for an enclosed bed. Good ventilation is important to control humidity and optimize the evaporation rate. As expected, evaporation occurs rapidly in warm, dry weather. Adaptation of mechanical sludge-removal equipment to enclosed beds is more difficult than to open drying beds.

Mechanical removal of sludge from drying beds has been practiced for many years at some large treatment plants, but now it is receiving more attention as the need increases to minimize problems with labor costs. Mechanical devices can remove sludges of 20%–30% solids while

TABLE 4-1
Design Criteria for Sludge Drying Beds

Type of digested sludge	Area (sq ft/capita)	Sludge loading dry solids (lb/sq ft/yr)
Primary	1.0	27.5
Primary and standard trickling filter	1.6	22.0
Primary and activated	3.0	15.0
Chemically precipitated	2.0	22.0

cakes of 30%–40% are generally required for hand removal. Small utility tractors with front-end loaders are often used for mechanical removal.

Drying times typically range from 4 to 12 weeks, depending on the weather. Especially adverse weather can result in drying times as long as 6 months.[14]

A major disadvantage in the larger plants likely to be involved in regional systems is the space required. For a 10-million-gallon-per-day activated-sludge plant, about 11 acres of drying beds would be required for the primary and WAS at a loading rate of 15 pounds per year per square foot. The space requirements, plus dependency on uncontrollable weather factors, are severe restrictions on the use of drying beds in large plants.

Figures 4-1* through 4-4 and Table 4-2 summarize costs for sand drying beds.

VACUUM FILTRATION

A vacuum filter basically consists of a cylindrical drum (see Figure 4-5) which rotates partially submerged in a vat of sludge. The filter drum is divided into compartments by partitions or seal strips. A vacuum is applied between the drum deck and the filter medium, causing filtrate to be extracted and filter cake to be retained on the medium during the pickup and cake-drying cycle. In the drum filter shown in Figure 4-5

---
*Illustrations for this chapter start on page 85.

TABLE 4-2
Estimated Construction Costs of Sand Drying Beds (Sept. 1976)

Cost component	Area (sq ft)			
	10,000	25,000	50,000	100,000
Concrete	800	$ 1,600	$ 3,200	$ 6,400
Steel	600	1,200	2,400	4,800
Metal piping and valves	4,000	9,300	18,600	37,200
Clay piping	1,800	4,200	8,400	16,800
Media	1,400	3,500	7,000	14,000
Labor	20,000	43,200	84,500	162,400
Miscellaneous items	4,300	9,500	18,600	36,200
Contingency	4,900	10,900	21,400	41,700
Total Estimated Cost	$37,800	$83,400	$164,100	$319,500

the cake of dewatered sludge is removed by a fixed scraper blade. There are alternative designs which use other methods for sludge removal.

The performance of vacuum filters may be measured by various criteria, such as the yield, the efficiency of solids removal, and the cake characteristics. Each of these criteria is of importance, but one or the other may be particularly significant in a given plant. Typical results are shown in Table 4-3.

Yield is the most common measure of filter performance. The yield is the filter output and is expressed in terms of pounds of dry total solids in the cake discharged from the filter, per square foot of effective filter area, per hour.

The second measure of filter performance is the efficiency of solids removal. Basically, the vacuum filter is a device used for separating solid matter from liquid, and the actual efficiency of the process is the percentage of feed solids recovered in the filter cake. Solids removals on vacuum filters range from about 85% for coarse-mesh media to 99% with close-weave, long-nap media. The recycled filtrate solids impose a load on the plant treatment units, and should normally be kept to a practical minimum. It may be necessary, however, to reduce the filter efficiency in order to deliver more filter output, and thus keep up with sludge production.

TABLE 4-3
Typical Results of Vacuum Filtration

Sludge type	Design assumptions	Percent solids to VF	Typical loading rates, (psf/hr)	Percent solids VF cake
Primary	Thickened to 10% solids, polymer conditioned	10	8–10	25–38
Primary + $FeCl_3$	85 mg/l $FeCl_3$ dose, lime conditioning, thickening to 2.5% solids	2.5	1.0–2.0	15–20
Primary + low lime	300 mg/l lime dose, polymer conditioned, thickened to 15% solids	15	6	32–35
Primary + high lime	600 mg/l lime dose, polymer conditioned, thickened to 15% solids	15	10	28–32
Primary + WAS	Thickened to 8% solids, polymer conditioned	8	4–5	16–25
Primary + (WAS + $FeCl_3$)	Thickened to 8% solids, $FeCl_3$ & lime conditioned	8	3	20
(Primary + $FeCl_3$) + WAS	Thickened primary sludge to 2.5%, flotation thickened WAS to 5%, dewater blended sludges	3.5	1.5	15–20
Waste activated sludge (WAS)	Thickened to 5% solids, polymer conditioned	5	2.5–3.5	15
WAS + $FeCl_3$	Thickened to 5% solids, lime + $FeCl_3$ conditioned	5	1.5–2.0	15
Digested primary	Thickened to 8%–10% solids, polymer conditioned	8–10	7–8	25–38
Digested primary + WAS	Thickened to 6%–8% solids, polymer conditioned	6–8	3.5–6	14–22
Digested primary + (WAS + $FeCl_3$)	Thickened to 6%–8% solids, $FeCl_3$ + lime conditioned	6–8	2.5–3	16–18
Tertiary alum	Diatomaceous earth precoat	0.6–0.8	0.4	15–20

The filter cake quality is another measure of filter performance, depending on cake moisture and heat value. Cake solids content varies from 20% to 40% by weight, depending on the type of sludge handled and the filter cycle time and submergence. Delivery of a very dry cake does not necessarily indicate good filter performance. Cake moisture should be adjusted to the method of final disposal, and it is inefficient to dry the cake more than is required. When incineration is practiced, a raw sludge cake having a fairly high moisture content can be burned without auxiliary fuel because of the higher volatile content, while a digested sludge cake will have to be drier to burn successfully without makeup heat. One approach to improving the filtration and incineration characteristics of primary-WAS mixtures is to feed powdered coal as a conditioning agent prior to the dewatering step.[33] It was found that about 0.3 pounds of coal per pound of dry sludge solids produced a sludge cake which permitted autogenous combustion with no effect on filter yield.

The effect of heat treatment prior to vacuum filtration on various municipal sludges is to make all types dewaterable to approximately the same degree.[34] Heat treatment provides a sludge that is readily dewaterable from primary or secondary sludges. Raw primary sludges have been dewatered at rates as high as 40 pounds per square foot per hour and waste-activated sludges at 7 pounds per square foot per hour. Mixtures of raw primary secondary sludges subjected to heat treatment should produce yields well over 10 pounds per square foot per hour.

Construction costs of vacuum filters and associated equipment and structures are most closely related to the filter surface area. Filter surface area requirements are in turn related to the type, quantity, and quality of sludge to be dewatered, sludge-conditioning processes used, and average hours of filter operation. Construction costs are related to respective filter surface areas in Figure 4–6 and Table 4–4. Estimated construction costs include the costs of the vacuum filters, auxiliary equipment, piping, and structures. Labor requirements for operating sludge filters are principally related to the size of the filters and the duration and frequency of filter runs. Operating labor requirements include filter start-up time, operation of the filters, and cleanup after the filter run. Also included in Figure 4–7 are operation of sludge-pumping and conditioning facilities prior to treatment and conveyor operation. Maintenance labor includes all labor related to maintenance of sludge filters and associated facilities.

Total costs of vacuum filtration typically range from $25 to $50 per ton of dry solids.

TABLE 4-4
Estimated Construction Costs of Vacuum Filtration (Sept. 1976)

	Filter area (sq ft)					
Cost component	113	200	300	400	600	760
Manufactured equipment	$ 93,100	$126,400	$146,400	$193,200	$232,800	$266,100
Labor	32,900	44,600	51,700	68,200	82,200	93,900
Housing	9,600	18,400	28,700	40,200	55,300	69,200
Metal piping and valves	20,000	31,700	37,800	45,000	55,000	80,000
Electrical	31,100	44,200	52,900	69,300	85,100	101,800
Miscellaneous items	28,000	39,800	47,600	62,400	76,600	91,700
Contingency	32,200	45,800	54,800	71,700	88,100	105,400
Total Estimated Cost	$246,900	$350,900	$419,900	$550,000	$675,100	$808,100

CENTRIFUGATION

There are many types of centrifugal equipment available, for a variety of specialized applications in industry.[35, 36] The solid-bowl centrifuge is the most widely used type for dewatering sewage sludge, however. The solid-bowl/conveyor sludge-dewatering centrifuge assembly (Figure 4-10) consists of a rotating unit comprising a bowl and conveyor. The solid cylindrical-conical bowl, or shell, is supported between two sets of bearings and includes a conical section at one end to form a dewatering beach or drainage deck over which the helical conveyor screw pushes the sludge solids to outlet ports and then to a sludge-cake discharge hopper. Sludge slurry enters the rotating bowl through a stationary feedpipe extending into the hollow shaft of the rotating screw conveyor, and is distributed through ports into a pool within the rotating bowl.

As the liquid sludge flows through the cylindrical section toward the overflow devices, progressively finer solids are settled centrifugally to the rotating bowl wall. The helical rotating conveyor pushes the solids to the conical section where the solids are forced out of the water, and free water drains from the solids back into the pool.

There are several variables which affect the performance of solid-bowl centrifuges. Bowl speed is one of the prime variables since centrifugal force speeds up the separation of solids from liquids. At any given pool depth, an increase in bowl speed provides more gravity-settling force, providing greater clarification. Typical g values for a solid bowl machine for many years were about 3,000. In recent years, units which operate at $g = 700$ have been developed. These "low"-speed units provide comparable results at lower power consumption.

The introduction of polymers has increased the range of materials that can be dewatered satisfactorily by centrifuges. The degree of solids recovery can be regulated over rather wide ranges depending on the amount of coagulating chemical used. Wetter sludge cake usually results from the use of flocculation aids because of the increased capture of fines.

Table 4-5 presents data on typical results with solid-bowl centrifugation. Heat-treated sludges will dewater to 35%–45% solids with no polymer required for 85% capture. Recovery of 92%–99% of the solids from heat-treated (primary) sludges has been reported with polymer dosage of 2–5 pounds per ton of dry solids.[34] Dewatering heat-treated mixtures of activated sludge and raw primary sludge has produced cake solids of 40% at 95% recovery without chemicals. Dewatering heat-treated activated sludges alone has achieved 35% cake solids at 95% recovery without chemicals. The use of 4 pounds per ton of polymers in this latter case enabled a 50% increase in centrifuge capacity while producing cake solids of 28%.[34]

TABLE 4-5
Typical Solid-Bowl Centrifuge Performance

Wastewater sludge type	Sludge cake characteristics		
	Solids (%)	Solids recovery (%)	Polymer addition
Raw or digested primary	28–35	70–90	no
Raw or digested primary, plus trickling filter humus	20–30	80–95	yes (5–15 lbs/ton)
		60–75	no
Raw or digested primary, plus activated sludge	15–30	80–95	yes (5–20 lbs/ton)
		50–65	no
Activated sludge	8–9	80–85	5–10 lbs/ton
Oxygen-activated sludges	8–10	80–85	3–5 lbs/ton
High-lime sludges	50–55	90	no
Lime classification	40	70	no

In addition to dewatering sludges, centrifuges have been used to separate impurities ("classify") from the lime sludges resulting from some phosphorus-removal processes to enable efficient recovery and reuse of the lime.

Centrifuges have the advantages of being a totally enclosed process and requiring less space than vacuum filters. They have the disadvantage of requiring more, and more highly skilled, maintenance.

A number of variables, including sludge-feed rate, solids characteristics, temperature, and conditioning processes, influence the selection of centrifugation equipment applicable to particular treatment plants. Of these, sludge-feed rate is the single parameter having the greatest influence on the equipment selection, and hence on construction costs.

The estimated relationship of centrifugation construction costs to installed capacity is shown in Figure 4–11 and Table 4–6. The cost estimates include allowances for centrifugation equipment, sludge-cake conveyors, equipment hoists, electrical facilities, and an enclosing structure.

The estimated labor requirements for operation and maintenance of sludge centrifugation and associated facilities are shown in Figure 4–12. The estimated annual payroll hours are related to the total annual dry solids in the sludge applied to the centrifuge. Included are labor requirements directly related to the centrifuge, sludge condi-

TABLE 4–6
Construction Costs for Centrifugation (Sept. 1976)[a]

Cost component	Firm capacity (gpm)				
	12	50	100	200	400
Manufactured equipment	$ 70,000	$161,500	$200,000	$300,000	$425,000
Labor	31,500	56,880	70,000	105,000	148,800
Housing	21,600	32,000	50,000	60,500	69,200
Electrical	21,500	39,710	48,000	69,800	96,400
Miscellaneous items	24,680	43,400	44,200	80,400	111,000
Contingency	28,380	50,000	63,400	92,400	127,600
Total Estimated Cost	$197,600	$383,490	$486,600	$708,100	$978,000

[a] Costs based on two machines minimum.

tioning, and other associated facilities. The energy requirements are based on $g = 700$ for dewatering. Typical costs are \$20–\$40 per ton of dry solids.

PRESSURE FILTRATION

During the last 5 years, there has been a substantial increase in use of pressure-filtration systems in U.S. wastewater plants. Improvements in the equipment involved, coupled with increasing quantities of difficult-to-dewater sludges, account for the increase.

The filter press is a batch device, which has been used in industry and in European wastewater plants for many years to process difficult-to-dewater sludges. There are several variations in mechanical design and operating pressures. For purposes of illustrating the concept, a vertical-plate filter-press system will be described. Such a press consists of vertical plates which are held rigidly in a frame and which are pressed together between a fixed and moving end as illustrated in Figure 4–15. On the face of each individual plate a filter cloth is mounted. The sludge is fed into the press and passes through the cloth, while the solids are retained and form a cake on the surface of the cloth. Sludge feeding occurs at pressures up to 225 psi and is stopped when the cavities or chambers between the trays are completely filled. Drainage ports are provided at the bottom of each press chamber. The filtrate is collected in these, taken to the end of the press, and discharged to a common drain. At the commencement of a processing cycle, the drainage from a large press can be in the order of 2,000–3,000 gallons per hour. This rate falls rapidly to about 500 gallons per hour as the cake begins formation, and when the cake completely fills the chamber the rate is virtually nothing. The dewatering step is completed when the filtrate is near zero. At this point the pump feeding sludge to the press is stopped and any back pressure in the piping is released through a bypass valve. The electrical closing gear is then operated to open the press. The individual plates are next moved in turn over the gap between the plates and the moving end. This allows the filter cakes to fall out. The plate-moving step can be either manual or automatic. When all the plates have been moved and the cakes released, the complete pack of plates is then pushed back by the moving end and closed by the electrical closing gear. The valve to the press is then opened, the sludge-feed pump started, and the next dewatering cycle commences.

Filter presses are normally installed well above floor level so that the cakes can drop onto conveyors or trailers positioned underneath the press. The pressures which may be applied to a sludge for removal of water by the filter presses now available range from 5,000 to 20,000 times the force of gravity. In comparison, a solid-bowl centrifuge provides forces of 700–3,500 g, and a vacuum filter, 1,000 g. As a result of these greater pressures, filter presses may provide higher cake-solids concentrations (30%–50% solids) at reduced chemical dosages. In some cases, ash from a downstream incinerator is recycled as a sludge conditioner.

Table 4–7 presents typical results from pressure filtration. As is

TABLE 4–7
Typical Results of Pressure Filtration

Sludge type	Conditioning	Percent solids to pressure filter	Typical cycle length (hrs)	Percent solids filter cake
Primary	5% $FeCl_3$, 10% lime	5	2	45
	100% ash		1.5	
Primary + $FeCl_3$	10% lime	4[a]	4	40
Primary + 2-stage high lime	None	7.5	1.5	50
Primary + WAS	5% $FeCl_3$, 10% lime	8[a]	2.5	45
	150% ash		2.0	50
Primary + (WAS + $FeCl_3$)	5% $FeCl_3$, 10% lime	8[a]	3	45
(Primary + $FeCl_3$) + WAS	10% lime	3.5[a]	4	40
WAS	7.5% $FeCl_3$, 15% lime	5[a]	2.5	45
	250% ash		2.0	50
WAS + $FeCl_3$	5% $FeCl_3$, 10% lime	5[a]	3.5	45
Digested primary	6% $FeCl_3$, 30% lime	8	2	40
Digested primary + WAS	5% $FeCl_3$, 10% lime	6–8[a]	2	45
	100% ash		1.5	50
Digested primary + (WAS + $FeCl_3$)	5% $FeCl_3$, 10% lime	6–8[a]	3	40
Tertiary alum	10% lime	4[a]	6	35
Tertiary low lime	None	8[a]	1.5	55

[a] Thickening used to achieve this solids concentration.

readily apparent, the process produces a drier cake than either vacuum filtration or centrifugation.

Figures 4–16 through 4–19 and Table 4–8 present cost information.

Labor requirements (Figure 4–17) are a function of the number of presses as well as the total cubic feet of press volume. A minimum level of effort appears to be 20 hours per 24 hours of operation of the filter press system for O & M of the press and related auxiliaries.

The bulk of the power consumed by the filter press (Figure 4–18) is the feed pump. The open and close mechanisms and tray-moving mechanism requirements are based on 2.5-hours-per-day operation of these components (20 hours per day of filtration time based on a 2-hour cycle, 0.4-hour turn-around time between cycles). The power requirements are a function of influent solids and the curve reflects 4%, 6%, and 8% influent solids.

Typical total costs for pressure filtration are $40–$75 per ton of dry solids. Although the dewatering costs are higher than for vacuum filters or centrifuges under comparable conditions, the drier cake produced may result in savings in the downstream process that are more than adequate to offset the higher costs.

DRYING LAGOONS

Lagoon drying is a low-cost, simple system for sludge dewatering that has been commonly used in the U.S. Drying lagoons are similar to

TABLE 4–8
Construction Costs for Pressure Filtration

Cost component	Volume (cu ft)				
	45	100	500	1,000	2,000
Manufactured equipment	$234,400	$469,000	$520,000	$1,040,000	$2,080,000
Labor	58,600	117,000	130,000	260,000	520,000
Electrical	45,900	72,100	97,500	195,000	390,000
Miscellaneous items	52,785	82,900	112,100	224,300	448,600
Contingency	60,700	95,400	128,900	257,900	515,800
Total Estimated Cost	$465,385	$836,400	$988,500	$1,977,200	$3,954,400

sandbeds in that the sludge is periodically removed and the lagoon refilled. Lagoons have seldom been used where the sludge is never removed, because such systems are limited in application to areas where large quantities of cheap land are available. Sludge is stabilized to reduce odor problems prior to dewatering in a drying lagoon. Odor problems can be greater than with sandbeds, because sludge in a lagoon retains more water for a longer period than does sludge on a conventional sand drying bed.

Other factors affecting design include consideration of groundwater protection and access control. Major design factors include climate, subsoil permeability, lagoon depth, loading rates, and sludge characteristics.

Solids loading rates suggested for drying lagoons are 2.2–2.4 pounds per year per cubic foot of lagoon capacity.[6] Other recommendations range from 1 square foot per capita for primary digested sludges in an arid climate to as high as 3–4 square feet per capita for activated-sludge plants where the annual rainfall is 36 inches. A dike height of about 2 feet with the depth of sludge (after decanting) of 15 inches has been used. Sludge depths of 2.5–4 feet may be used in warmer climates where longer drying periods are possible.

Sludge will generally not dewater in any reasonable period of time to the point that it can be lifted by a fork except in an extremely hot, arid climate. If sludge is placed in depths of 15 inches or less, it may be removed with a front-end loader in 3–5 months. When sludge is to be used for soil conditioning, it may be desirable to stockpile it for added drying before use. One proposed approach utilizes a 3-year cycle in which the lagoon is loaded for 1 year, dries for 18 months, is cleaned, and then allowed to rest for 6 months. Definitive data on lagoon drying are scarce. Sludge may be dewatered from 5% solids to 40%–45% solids in 2–3 years using sludge depths of 2–4 feet.

Figures 4-20 through 4-23 and Table 4-9 present cost information on drying lagoons.

Operation and maintenance costs are associated principally with removal of accumulated sludge. The energy and labor requirements were computed assuming sludge removal would be accomplished from one-half of the storage lagoons once per year and disposal would be within 1 mile of the lagoon site. If longer hauling distances are required, curves for sludge hauling should be utilized. Energy costs are derived from estimated diesel fuel consumption of 5 gallons per hour for removal and transport of the sludge to the disposal area. Maintenance materials are related only to costs of lagoon maintenance, and

TABLE 4-9
Estimated Construction Costs of Drying Lagoons (Sept. 1976)

Cost component	Volume (acre ft)				
	15	30	50	100	150
Earthwork	$15,000	$25,000	$38,000	$ 70,000	$100,000
Paving	5,000	7,000	9,000	13,000	15,000
Seeding	4,000	5,000	8,000	12,000	15,000
Miscellaneous items	4,000	6,000	8,000	14,000	20,000
Contingency	4,000	6,000	9,000	16,000	23,000
Total Estimated Cost	$32,000	$49,000	$72,000	$125,000	$173,000

include lagoon grading and restoration of dikes, grass cutting, roadway maintenance, etc. Labor costs were developed from estimated man-hour requirements for sludge removal, sludge transport, and facility maintenance.

BELT FILTER PRESS

The belt filter press combines features of the vacuum filter and the pressure filter, and under certain conditions offers advantages over these two methods of sludge dewatering.[37] There are several different configurations of equipment, but all combine the operations of drainage, pressure filtration, and shear-pressure filtration. Application of belt filter presses has achieved 35%–40% solids from polymer-conditioned primary sludge, and 28%–31% from a 50:50 mixture of primary and waste-activated sludge with solids recoveries of 97%–99%.

Costs are presented in Figures 4–24 through 4–27 and Table 4–10.

The energy requirements were developed from total connected horsepower for the belt drive unit, belt wash pump, conditioning tank, feed pump, polymer pump and tanks, belt conveyor, and electrical control panel. With the exception of the smallest machine, units were selected on the basis of 22 hours of continuous operation with 2 hours of down time for routine maintenance and a loading rate of 15 gallons per minute of sludge per meter of machine width.

TABLE 4-10
Estimated Construction Costs for Belt Filter Presses (Sept. 1976)

	Sludge flow (gpm)			
	2.5	25	250	500
Size machine belt width (m)	1	2	3	3
No. machines	1	1	5	10
Manufactured equipment	$ 80,000	$135,000	$ 850,000	$1,650,000
Installation labor	28,000	47,250	253,750	577,500
Housing	18,000	25,000	72,900	121,500
Electrical	18,900	31,088	176,498	352,350
Miscellaneous	21,735	35,751	202,972	405,203
Contingency	24,995	41,113	233,418	465,983
Total Estimated Cost	$191,630	$315,202	$1,789,538	$3,572,536

Sludge Dewatering/85

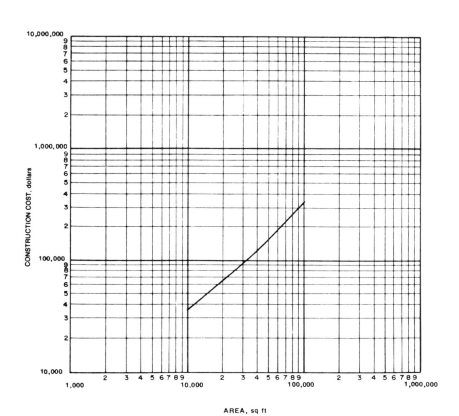

Figure 4-1. Construction costs for sand drying beds (Sept. 1976).

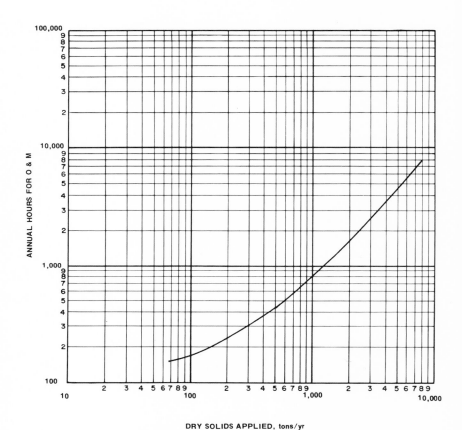

Figure 4-2. Labor requirements for sand drying beds.

Sludge Dewatering/87

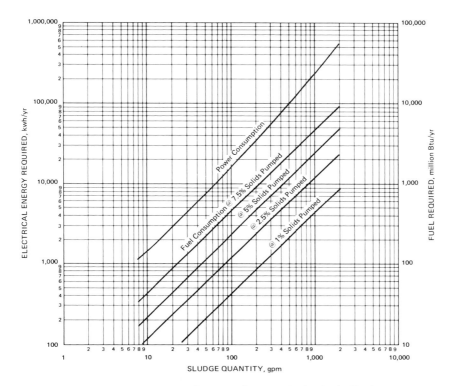

Figure 4-3. Energy requirements for sand drying beds. Design assumptions: power consumption is based on pumping to drying beds at TDH = 15 ft; fuel consumption is based on (1) drying to 50% solids, 70 lbs/cu ft; (2) loading with front-end loader, 8 gal/hr use of diesel fuel (140,000 BTU/gal); (3) 15 min required to load 30-cu-yd truck.

88/Handbook of Sludge-Handling Processes

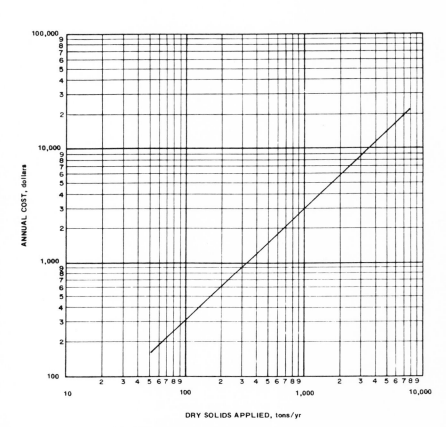

Figure 4-4. Maintenance material and supply costs for sand drying beds (Sept. 1976).

Sludge Dewatering/89

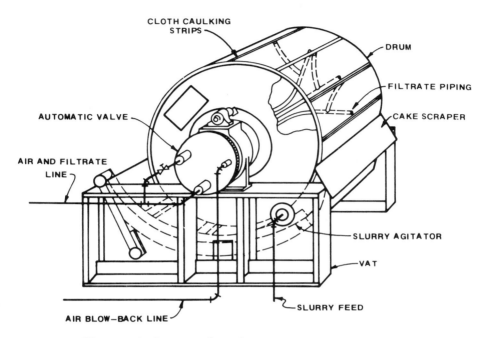

Figure 4–5. Cutaway view of a rotary drum vacuum filter (reference 6).

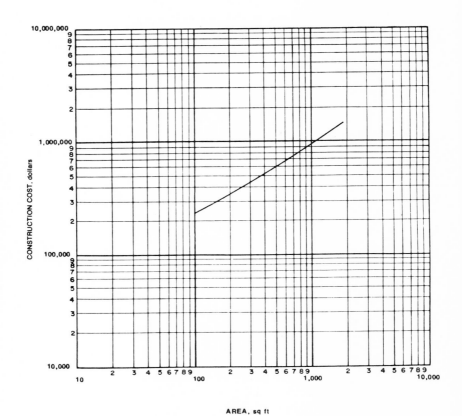

Figure 4-6. Construction costs for vacuum filtration (Sept. 1976).

Sludge Dewatering/91

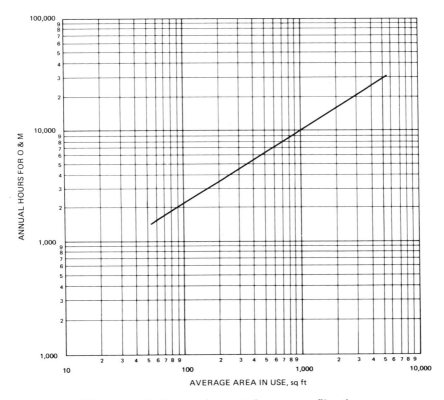

Figure 4-7. Labor requirements for vacuum filtration.

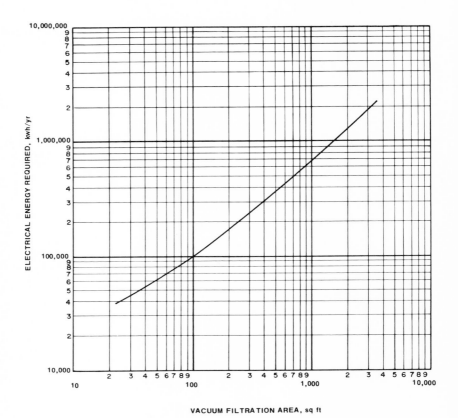

Figure 4-8. Energy requirements for vacuum filtration. Operating parameters: 2 scfm/sq ft; 20-22 inches Hg vacuum; filtrate pump, 50 ft TDH; curve includes: drum drive, discharge roller, vat agitator, vacuum pump, filtrate pump.

Sludge Dewatering/93

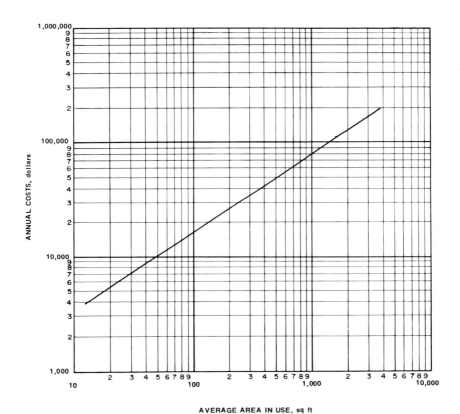

Figure 4-9. Maintenance material costs for vacuum filtration (Sept. 1976).

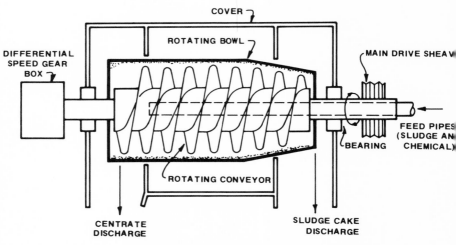

Figure 4-10. Continuous countercurrent solid-bowl conveyor discharge centrifuge.

Sludge Dewatering/95

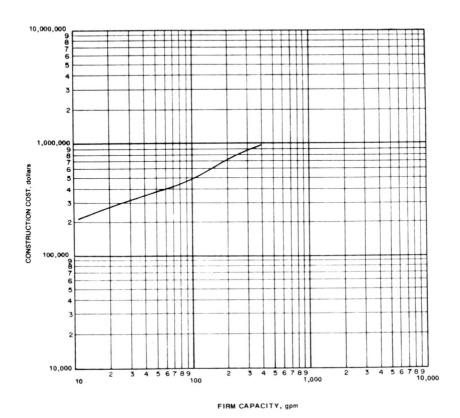

Figure 4–11. Construction costs for centrifugation (Sept. 1976).

96/Handbook of Sludge-Handling Processes

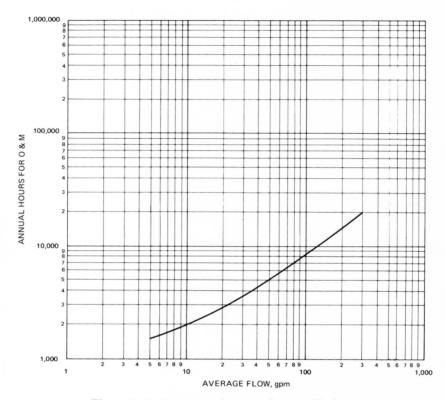

Figure 4-12. Labor requirements for centrifuging.

Sludge Dewatering/97

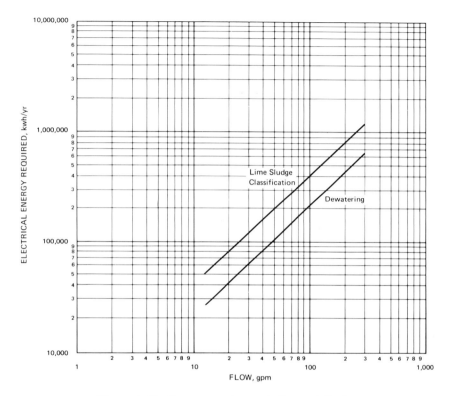

Figure 4-13. Energy requirements for centrifuging.

98/Handbook of Sludge-Handling Processes

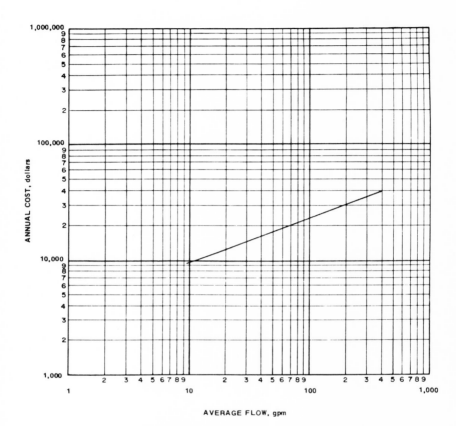

Figure 4-14. Maintenance material and supply costs for centrifugation (Sept. 1976).

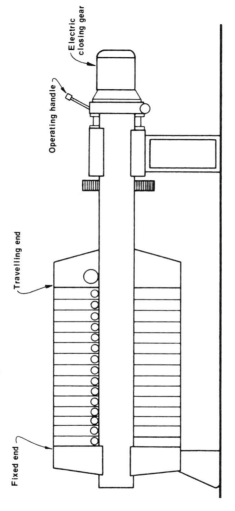

Figure 4-15. Side view of a filter press (reference 6).

100/Handbook of Sludge-Handling Processes

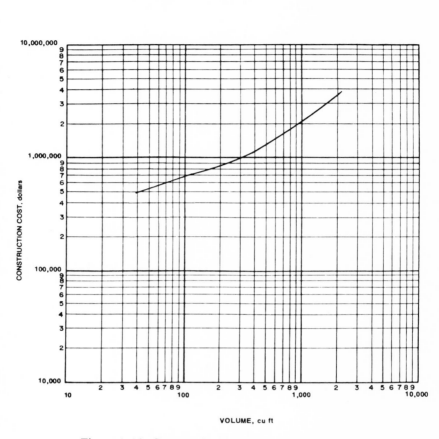

Figure 4-16. Construction costs for pressure filtration.

Sludge Dewatering/101

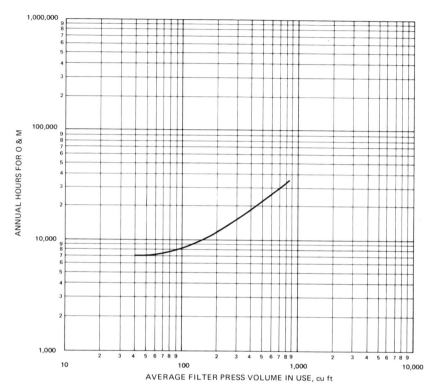

Figure 4–17. Labor requirements for pressure filtration (based on continuous, 7-day-week operation, 2-hr cycle).

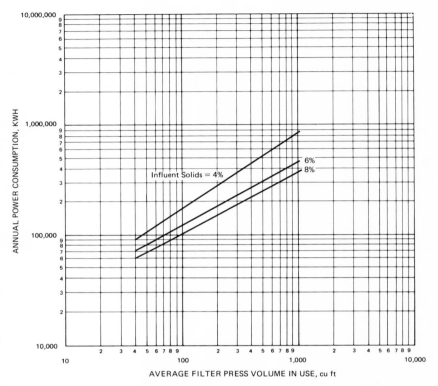

Figure 4–18. Power requirements for pressure filtration (based on a continuous, 7-day-week operation, 2-hr cycle).

Sludge Dewatering/103

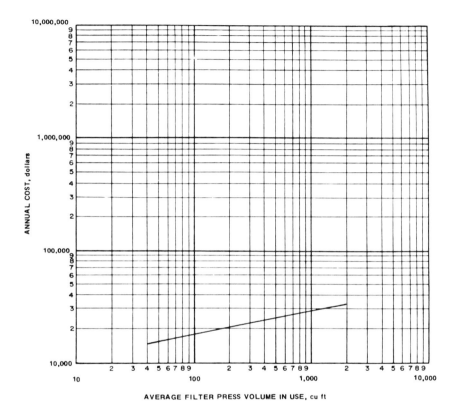

Figure 4-19. Maintenance material and supply costs for pressure filtration.

104/Handbook of Sludge-Handling Processes

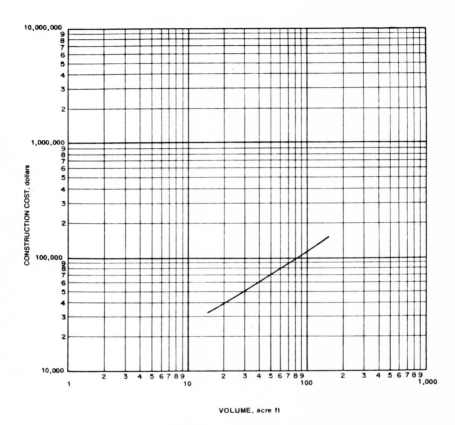

Figure 4-20. Construction costs for drying lagoons (Sept. 1976).

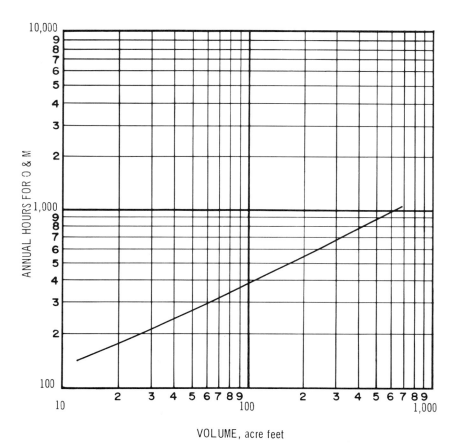

Figure 4-21. Labor requirements for drying lagoons.

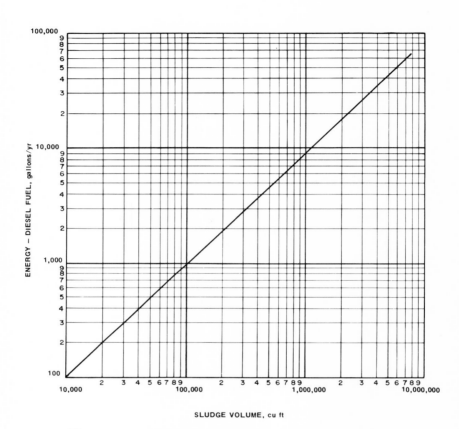

Figure 4-22. Energy requirements for sludge dewatering lagoons.

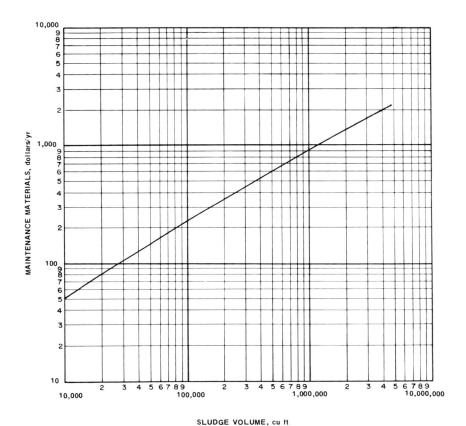

Figure 4-23. Maintenance material costs for sludge dewatering lagoons.

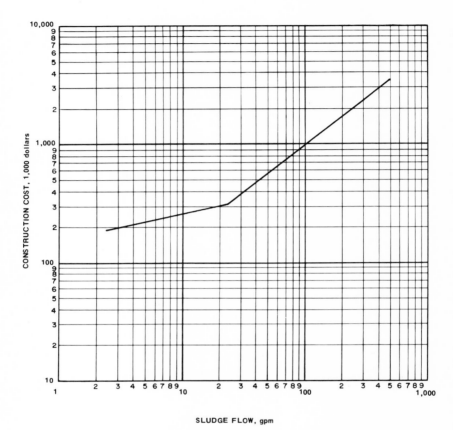

Figure 4-24. Construction costs for belt filter press (Sept. 1976).

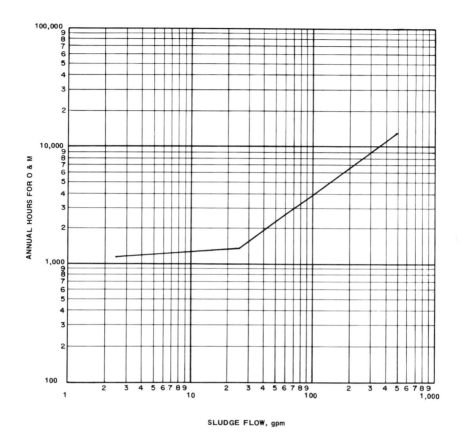

Figure 4-25. Labor requirements for belt filter press.

110/Handbook of Sludge-Handling Processes

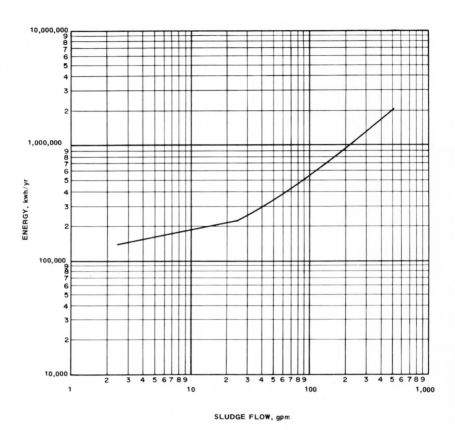

Figure 4–26. Energy requirements for belt filter press.

Sludge Dewatering/111

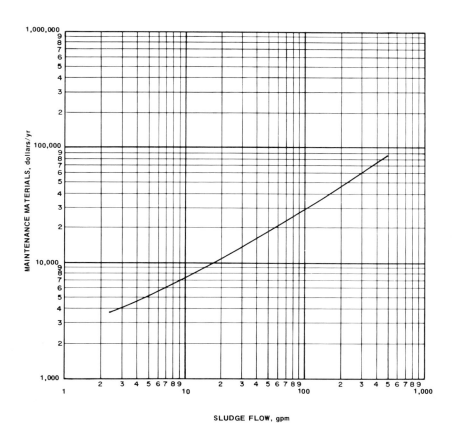

Figure 4-27. Maintenance materials costs for belt filter press.

Chapter 5

Sludge Stabilization

The principle purpose of sludge stabilization is to render the sludge less putrescible, to reduce the pathogenic content, and to reduce the sludge quantity. Stabilization processes include anaerobic and aerobic digestion, and composting.

ANAEROBIC DIGESTION

In this process, the organic matter in the sludge is stabilized in an anaerobic (oxygen-devoid) environment. Most modern systems are "high-rate" systems utilizing one or two stages (see reference 38 for detailed information on the process). In a typical two-stage process (Figure 5-1*), the stabilization of the sludge occurs in the first-stage, mixing and heating unit, with the second-stage digester providing settling and thickening. In a single-stage system, the secondary digester is replaced by some other thickening process. The digester is heated to 85°–95°F and typically provides 15 days or less detention of the sludge.

The process has been successful when primary sludge, or combinations of primary sludge and limited amounts of secondary sludge, constitute the system's feed. With the advent of wastewater-treatment systems that are more efficient than simple sedimentation, large quantities of activated sludges are produced at the plants. This additional sludge, when placed in a two-stage anaerobic digestion process, can cause high operating costs and poor plant efficiencies. The basic cause of the problem is that the additional solids do not readily settle or dewater after digestion. The process converts about 50% of the organic solids to liquid and gaseous forms, providing a substantial reduction in the quantity of sludge requiring disposal.

Digestion tank volume requirements, to which construction costs are related, depend on a number of factors including the quantity and

*Illustrations for this chapter start on page 127.

characteristics of sludge, digester operating temperature, digested-sludge storage requirements, and the degree of digestion desired. A control building enclosing sludge circulating, heating, and control facilities normally has walls in common with adjacent digesters. A frequent configuration includes a control building located between a pair of digesters.

Construction costs are plotted against respective sludge volumes in Figure 5-2. The volumes used are computed on the basis of the side-water depth only, and exclude the freeboard and conical bottom volumes.

The construction cost per unit of digester volume decreases as total volume increases up to the maximum volume of two digesters with a control building. The largest volume practical in such an installation is about 400,000 cubic feet, based on two digesters with diameters of 100 feet and side-water depths of about 25 feet. Construction cost estimates shown in Figure 5-2 for volumes larger than 400,000 cubic feet are based on the assumption that multiples of this arrangement would be utilized without further reduction in cost per unit of digester volume.

The estimates of costs shown in Table 5-1 include allowances for normal digester and control-building structures, sludge heating, circulating and control equipment, and piping within the limits of the structures. Cost of piping to and from the digesters is not included in these estimates.

Operation and maintenance costs presented in Figures 5-3 through 5-5 include the digestion tanks, control building, and equipment used in the digestion of sludge. The tanks are frequently equipped with floating covers and mixers. The associated control building usually has sludge heat exchangers and control facilities, and often includes circulation pumping equipment.

Operation labor is required to monitor and control the sludge-digestion facilities. Various tests and measurements are made at these facilities and the results recorded to indicate to the operator when adjustments or changes should be made in the digestion process, including when to remove digested sludge and supernatant from the digesters. Sludge is added to digesters at frequent intervals and the proper temperature is maintained by circulating the digester contents through the sludge heaters. This plant component requires proper and adequate operating labor if these facilities are to be efficiently operated.

Maintenance labor is required at digesters to clean, maintain, and repair the structures and equipment used in the sludge digestion process. Items requiring the most maintenance include floating digester

TABLE 5-1
Estimated Construction Cost of Anaerobic Digesters (Sept. 1976)

Cost component	Volume (1,000 cu ft)					
	5	25	100	200	700	1,500
Manufactured equipment	$ 28,000	$ 65,200	$157,100	$247,000	$654,300	$1,557,900
Concrete	4,300	9,900	23,600	36,200	95,900	228,300
Steel	4,700	11,000	26,600	40,600	107,500	256,000
Labor	28,700	56,500	134,400	176,500	467,500	1,113,100
Metal pipe and valves	10,000	12,500	15,000	21,000	55,600	132,400
Electrical	11,400	23,300	53,500	78,200	207,100	493,200
Miscellaneous items	13,100	26,800	61,500	89,900	238,200	567,100
Contingency	15,000	30,800	70,800	103,400	273,900	652,000
Total Estimated Cost	$115,200	$236,000	$542,500	$792,800	$2,100,000	$5,000,000

covers, gas-recirculation equipment, mechanical mixers, heat exchanger, pumps, control equipment, and gauges.

Heat is required in the anaerobic digestion process (1) to raise the temperature of the influent sludge to the level of the digester, and (2) to compensate for heat losses from the digester through its walls, bottom, and cover.

The heat required to raise the influent sludge temperature can be calculated from the following relationship:

$Q = WC(T_D - T_S)$
Q = heat required in BTUs
W = weight of influent sludge in pounds
C = specific heat of sludge, 1.0 BTU per pound per °F for 1%–10% solids sludge
T_D = temperature in digester in °F
T_S = temperature of influent sludge in °F

Based on loadings of 0.05 (standard rate) and 0.15 (high rate) pounds of volatile solids per day per cubic foot, the digester capacities shown in Table 5-2 result.

The total heat required for digestion at 95°F is shown in Figure 5-6 for primary sludge and Figure 5-7 for primary plus waste-activated sludge. These heat requirements are based on the above criteria for sludge heating and digester heat loss and 75% heat-transfer efficiency.

A major component of the gaseous by-products (usually about two-thirds) is methane. The resulting gas has a typical heat value of 600 BTUs per standard cubic foot with about 15 standard cubic feet of gas formed per pound of volatile solids destroyed.

The use of anaerobic digester gas has been practiced to some extent in wastewater-treatment plants for many years. Digester gas is currently being used at several wastewater-treatment plants to heat digesters and buildings, and as fuel for engines that drive pumps, air blowers, and electrical generators.

Table 5-3 gives estimates for gas and heat available from anaerobic digestion.[38]

A schematic of a typical system to utilize digester gas in an internal combustion (IC) engine is shown in Figure 5-8. (As indicated in this figure, the engine could be coupled to a generator, blower, or pump.) Typical IC engine efficiency is 36.4% (7,000 BTUs per horsepower-hour). An IC engine-generator's typical efficiency is 30% (11,400 BTUs per horsepower-hour). The electrical energy which can be generated from anaerobic digestion of primary sludge and WAS could supply

TABLE 5-2
Digester Capacities

Sludge type	Solids Content %	Total solids (lb/mil gal)	Volatile solids (lb/mil gal)	Total sludge (lb/mil gal)	Digester capacity (cu ft/mil gal) Loading (lb VS/day/cu ft)	
					0.05	0.15
Primary	5	1,155	690	23,100	13,800	4,600
Primary + WAS	4.5[a]	2,096	1,446	46,600	28,900	9,600

[a]Thickened.

TABLE 5-3
Gas and Heat Available from Anaerobic Digestion

Criterion	Primary sludge	Waste activated sludge	Total
Gas produced, scf per million gallons treated	5,175	5,670	10,845
Heat available, BTU per million gallons treated	3,105,000	3,402,000	6,507,000

about 85% of the electrical energy required for an activated-sludge plant while also providing over 50% of the heat for the digestion process itself.[39]

Estimated costs to clean and store digester gas (the first steps in Figure 5-8) are summarized in Table 5-4. Hydrogen sulfide can be removed from digester gas by treatment in a chemical scrubbing system using sodium hypochlorite or other oxidizing agents. Estimated costs include scrubbing with NaOCl in a packed tower to remove 1,000 ppm H_2S and on-site hypochlorite generation. It is possible to use activated carbon for H_2S removal, but the carbon must be regenerated with steam. Chemical scrubbing systems appear to be more economical and

TABLE 5-4
Digester Gas Cleaning and Storage Costs

Plant capacity (mgd)	Scrub and compress (scfm)	H_2S removed @ 1,000 ppm[a] (lb/day)	Gas compressed[b] (1,000 cu ft/day)	Scrub and compress construction cost ($1,000)
5	50	5	24	88
10	100	10	48	99
25	200	25	96	117
50	350	50	168	156
75	640	75	307	205
100	1050	100	504	257
200	1400	200	672	367
300	2100	300	1008	450

[a] Assumes digester gas = 0.071 lb/cu ft.
[b] Gas compressed and stored @ 45 psi.

simpler to operate. It may be possible to use other chemicals, or other sources of hypochlorite, to furnish less expensive scrubbing systems than shown in Table 5-4. Iron sponge scrubbers have been installed in some treatment plants. Construction costs for cleaning and storing digester gas are greatly influenced by the storage capacity provided. The storage capacity used in these estimates is based on one sphere per plant, up to plant sizes of 100 million gallons per day.

Estimated costs for 600-rpm IC engines equipped with heat-recovery and alternate-fuel systems are shown in Figures 5-9 and 5-10. These cost curves include data for both dual-fuel and gas engines. Operation and maintenance costs are greatly affected by the alternate fuel consumed. Propane alternate-fuel systems are more costly than fuel-oil systems; however, gas engines that require propane are less costly than dual-fuel engines that require fuel oil. Dual-fuel engines require about 10% fuel oil on an average annual basis. Gas engines could operate without using any alternate fuel. For these estimates, however, it is assumed that 10% propane would be consumed. Propane would have to be used (or at least paid for) to obtain contracts for a firm supply.

Estimated costs for complete systems to generate electricity with digester gas are shown in Figures 5-11 and 5-12. These costs are for a system as shown in Figure 5-8. The cost curves may be used to estimate on-site electricity generation costs as shown in Table 5-5 for a 100-

Storage spheres			Cleaning and storage system[c]			
Volume (1,000 cu ft)	Number diam. (ft)	Construction cost ($1,000)	Construction cost ($1,000)	Labor (hr/yr)	Material ($1,000/yr)	Energy (1,000 kwh/yr)
17	1/32	65	153	240	2	142
24	1/36	90	189	470	4	219
50	1/46	185	302	1000	10	371
74	1/36 1/46	275	431	2000	20	634
113	1/60	400	605	2900	30	1092
113	1/60	400	657	3750	40	1593
226	2/60	800	1168	3750	60	2533
339	3/60	1200	1650	7500	80	3800

[c] Cleaning and storage system includes scrubbers, compressors, and storage spheres in a complete system.

TABLE 5-5
Example of On-Site Electricity Generation Costs

Construction cost (Figure 5-11)	$2,500,000	
Material (Figure 5-11)	55,000	per year
Labor (Figure 5-12)	5,800	hr/yr
Electricity (Figure 5-12)	1,500,000	kwh/yr
Fuel (Figure 5-12)	23×10^9	BTU/yr
Annual costs:		
Construction ($2,500,000 plus 35% for engineering, administration, interest during construction, and other costs = $3,375,000 total; amortize for 20 years at 7% interest: $3,375,000 \times 0.09439 = $319,000)	$319,000	per year
Operation and Maintenance	$220,000	per year
Labor 5,800 hr @ $10/hr $58,000		
Material $55,000		
Electricity 1,500,000 kwh @ $0.025/kwh $38,000		
Fuel 23×10^9 BTU/yr @ 3/mil BTU $69,000		
Total Annual Cost	$539,000	per year

million-gallons-per-day plant with 2400 kilowatts per hour potentially available from digester gas, demonstrating a unit electricity production cost of $0.026 per kilowatt-hour. If the generating facility operates only 80% of the time, the unit cost increases to $0.032 per kilowatt-hour.

AEROBIC DIGESTION

Aerobic digestion consists of separate aeration of waste primary sludge, waste biological sludge, or a combination of waste primary and biological sludges in an open tank. It is usually used to stabilize excess activated sludges or the excess sludges from small plants which do not have separate primary clarification. Figure 5-13 is a schematic diagram of an aerobic digestion system. The advantages that the system offers over anaerobic digestion include simpler operation, less capital cost, and better supernatant quality. Its disadvantages are higher operating cost, poor sludge dewatering characteristics, and net energy consumption rather than energy production.

Current practice is to provide 10–15 days of detention time for the stabilization of excess biological sludges. Additional time is required when primary sludge is included.[6]

The destruction of solids is a function of temperature (the process is not heated). Volatile solids reductions of 35%–50% have been achieved. Pure oxygen rather than air can be used in the digester to achieve higher loading rates. The oxygen system also has the advantage of generating heat from the biological reaction which increases the sludge temperature and correspondingly increases the rate of solids destruction.

The capital costs for aerobic digestion shown in Figure 5–14 and Table 5–6 include a minimum of two tanks. Each tank includes floating aeration equipment, provisions for sludge feed and drawoff, and provisions for drawoff of supernatant. Floating aeration equipment is used because the tank level is allowed to vary. Controls are provided so that mixing is stopped for periods of time for settling and supernatant removal. A control structure is provided between the tankage similar to anaerobic digesters. Operation and maintenance costs (Figures 5–15 through 5–17) are most closely related to the horsepower of the aeration equipment. Capital costs for aerobic digestion are significantly less than for two-stage anaerobic digestion partly because total sludge detention time is about one-third that of two-stage anaerobic digestion. O & M

TABLE 5–6
Construction Costs for Aerobic Digestion (Sept. 1976)

Cost component	Volume (1,000 cu ft)				
	10	50	100	500	1,000
Manufactured equipment	$ 9,200	$ 20,000	$25,800	$ 374,900	$ 675,900
Labor	19,400	55,300	84,000	278,100	460,800
Concrete	2,600	14,000	23,000	87,200	155,000
Steel	8,400	25,700	39,200	127,500	212,100
Electrical	1,600	9,100	14,100	46,300	92,600
Miscellaneous items	4,400	13,700	21,300	71,600	121,200
Contingency	6,800	20,700	31,100	147,800	257,600
Total Estimated Cost	$52,400	$158,500	$238,500	$1,133,400	$1,975,200

costs are significantly higher for the aerobic process because of its greater power consumption. In order properly to compare these two systems, differences in thickening, dewatering, and supernatant-treatment costs must be added to each of the digestion-system costs.

COMPOSTING

Composting is a method of biological oxidation of organic matter in sludge by thermophilic organisms. Composting, properly carried out, will dewater, destroy objectionable odor-producing elements of sludge, destroy or reduce disease organisms because of the elevated temperature, and produce a useful organic product.

Composting wastewater sludge differs significantly from processing and composting solid waste; therefore, past poor publicity related to composting solid waste need not discourage the use of composting in processing wastewater sludge. There are many differences between the two:

> Composting solid waste proceeds by a complex materials-handling and separation process.
> Solid waste varies widely in composition, which makes processing more difficult.
> Many past solid-waste composting operations were operated and evaluated on the basis of profit-making potential rather than as an alternative disposal means.
> For a given population, the volume of solid-waste compost is several times the volume of wastewater-sludge compost; therefore, solid waste creates a much greater marketing or disposal task.

Composting systems generally fall into three categories: (a) pile, (b) windrow, and (c) mechanized or enclosed systems. The pile (static aerated pile) and windrow systems have been used almost exclusively in composting sewage sludge because of their low cost and demonstrated performance. In general, the windrow process has been used in composting digested primary and waste-activated sludge in various combinations. The static-pile method has been used more recently for composting raw primary and waste-activated sludge in various combinations. The windrow process was found to be unsuitable for composting raw sludge because of odor problems. Thus, at this time, the windrow process has been demonstrated on digested sludges, the static-pile method on raw sludges, and mechanized or enclosed systems have not been used to any extent recently in the U.S. on sewage sludge. Thus, only windrow and static-pile processes will be discussed in this section.

The general composting method is very similar for both processes. The dewatered sludge (typically 20% solids) is delivered to the site and is usually mixed with a bulking agent. The purpose of the bulking agent is to increase the porosity of the sludge to assure aerobic conditions during composting. If the composting material is too dense or wet it may become anaerobic, thus producing odors. If it is too porous, the temperature of the material will remain low. Low temperatures will delay the completion of composting and reduce the kill of disease organisms.

Various bulking materials can be used, and suitable low-cost materials include wood chips, bark chips, rice hulls, and cubed solid waste. Unscreened finished compost has also been used. Generally, one part sludge (20% solids) is mixed with three parts bulking agent, although this mixture can be varied depending on moisture content of sludge, type of bulking agent, and local conditions. The sludge–bulking agent mixture is then formed into the windrow or static pile as applicable.

Following composting, the product is removed from the windrow or static pile and cured in storage piles for 30 days or longer. This curing provides for further stabilization and pathogen destruction. Prior to or following curing, the compost may be screened to remove a portion of the bulking agent for reuse or for applications requiring a finer product. The compost can also be used without screening. Removal of the bulking agent also reduces the dilution of the nutrient value of the compost.

The compost is then ready for distribution.

Windrow Composting

The sludge–bulking agent mixture (two to three parts of bulking agent by volume to one part of sludge) is spread in windrows with a triangular cross section. The windrows are normally 10–16 feet wide and 3–5 feet high. An alternative method of mixing the bulking agent and sludge and forming the windrow consists of laying the bulking agent out as a base for the windrow. The sludge is dumped on top of the bulking agent and spread. A composting machine (similar to a large rototiller) then mixes the sludge and bulking agent and forms the mixture into a windrow. Several turnings (about 8–10 times) are necessary to blend the two materials adequately.

The windrow is normally turned daily using the composter; however, during rainy periods turning is suspended until the windrow surface layers dry out. Temperatures in the windrow interior under proper composting conditions range from 55° to 65°C. Turning moves the surface material to the center of the windrow for exposure to the higher temperatures needed for pasteurization to kill most pathogenic agents. Turning also aids in drying, and increases the porosity for greater air movement and distribution.

The windrows are turned for a two-week period or longer, depending on the weather and efficiency of composting. The compost windrow is then flattened for further drying. The compost is moved to curing when the moisture content has decreased to approximately 30% to 40%. Proper windrow composting should produce a relatively stable product with a moisture content of 30%–40% which has been exposed to temperatures of at least 50°C for a portion of the composting process.

The composting process requires longer detention times in cold or wet weather; therefore, climate is a significant factor with the windrow process in open spaces. Covering the composting area would significantly reduce the effects of cold weather and nearly eliminate the problems of wet weather. In any case, the curing area should be covered if operations are to be carried out during precipitation.

Static-Pile Composting

The static-pile composting method[41, 42] as applied to raw sludge requires a forced-ventilation system for control of the process. The pile then remains fixed, as opposed to the constant turning of the windrow, and the forced-ventilation system maintains aerobic conditions.

A base is prepared for the pile consisting of a 1-foot-thick layer of bulking agent or previously composted unscreened product. A 4-inch-diameter perforated pipe is installed in the base as an aeration header. The base is constructed with a typical plan dimension of approximately 40 by 20 feet. The sludge–bulking agent mixture is piled on this base to a height of approximately 8 feet to form a triangular cross section. The pile is capped with a 1-foot layer of screened compost product. This top layer extends down the sides to help absorb odors and to shield against penetration of precipitation. A typical static pile is illustrated in Figure 5–18. An alternative configuration is the extended static pile method, where subsequent piles are "added" to the initial static pile. This configuration saves space compared to a number of separate static piles.

The perforated underdrain pipe is attached to a blower by pipe and fittings. The other side of the blower is piped to an adjacent smaller pile of screened compost product. Air and gases are drawn by the blower from the static compost pile and discharged through the small pile of product compost. The small pile effectively absorbs odors. The operating cycle of the blower is adjusted to maintain oxygen levels in the exhausted gases and compost pile within a range of 5%–15%. Temperatures within the compost pile will vary somewhat with monitoring location in the pile, but should reach 60°–65°C. Normally the blower is operated on an on-off cycle to maintain proper oxygen levels and temperatures within the pile.

After an average composting period of three weeks, the compost is moved to the curing area. Land requirements are affected by several factors but are typically 0.2–0.4 acres per dry ton for the static-pile technique. Windrow techniques require two to three times more area.

Outdoor temperatures as low as $-7°C$ and rain totaling 7 inches per week has not interfered with the successful outdoor operation of exposed static-pile composting. Temperatures produced during static-pile composting are generally above 55°C and often exceed 70°–80°C.

The compost product has a slight musty odor, is moist and dark in color, and can be bagged. The texture of the compost varies depending on the degree of screening. Compost is valuable as a soil conditioner and low-grade fertilizer, and varies widely in content. Typical compost contains an average of 1.5% nitrogen and 1.0% phosphorus. Agricultural Research Service personnel indicate that proper static-pile composting should reduce total and fecal coliform and salmonella below detectable limits. Compost produced by the windrow process is likely to contain detectable pathogens because lower temperatures are produced. Composting has little effect on total heavy metal content of the sludge, but there is some dilution and also some indication of lower uptake rate after composting. Content and effect of heavy metals must be considered for each individual application.

The construction costs shown in Figure 5–19 for composting facilities are based on data from the few actual installations and on estimates extrapolated from actual and planned installations. The costs include site preparation, paving, aeration equipment, site office, and equipment as required for the forced-aeration static-pile method. Operation and maintenance requirements (Figures 5–20 through 5–22) include labor for all operations, including monitoring and runoff control. A depreciation allowance is included for periodic replacement of mobile equipment. Sludge and compost hauling is not included. Requirements were developed from actual forced-air static-pile and windrow operations.

The marketing of the end product is a key to the success of a composting effort. A recent market study found several successful municipal sludge-composting operations where all of the end product was sold or otherwise successfully used.[43] The study concluded that the upper price limit for bulk sewage sludge compost would be $4–$10 per ton, and for packaged, bagged sewage sludge compost, $60 per ton. Bagging costs could approach $30 per ton.

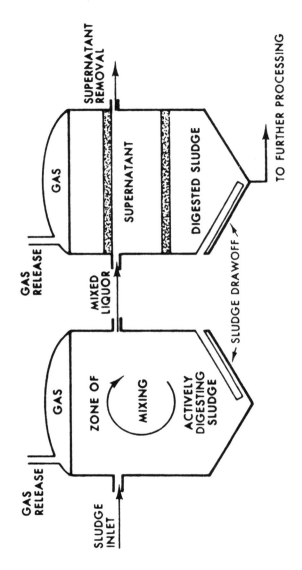

Figure 5–1. Two-stage anaerobic digestion.

128/Handbook of Sludge-Handling Processes

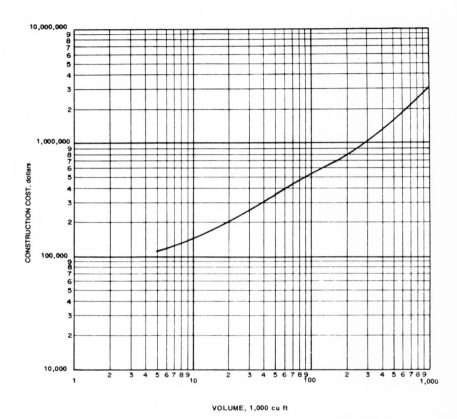

Figure 5-2. Construction costs for anaerobic digestion (Sept. 1976).

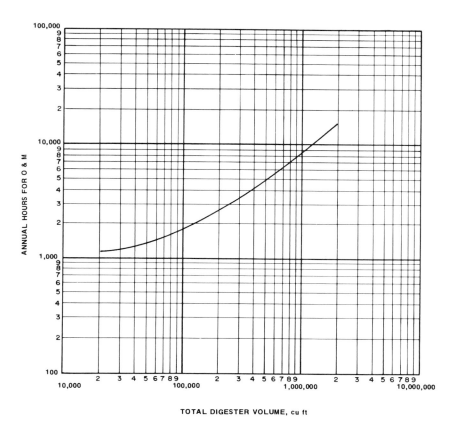

Figure 5-3. Labor requirements for anaerobic sludge digestion.

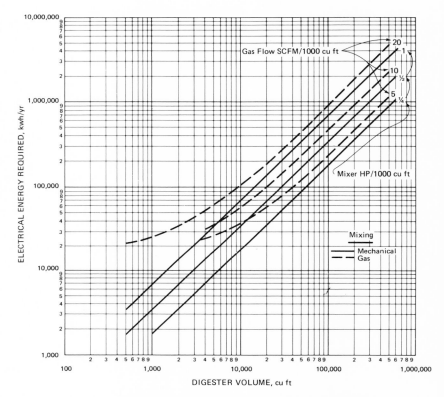

Figure 5-4. Energy requirements for anaerobic digester (high-rate mixing). Design assumptions: continuous operation; 20 ft submergence for release of gas; motor efficiency varies from 85% to 93% depending on motor size.

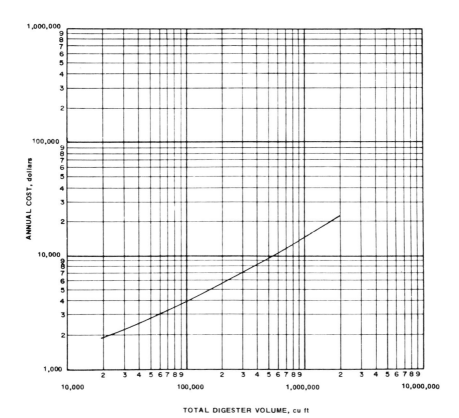

Figure 5-5. Maintenance material and supply costs for anaerobic sludge digestion (Sept. 1976).

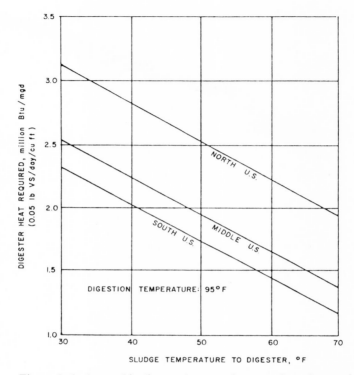

Figure 5-6. Anaerobic digester heat requirements for primary sludge.

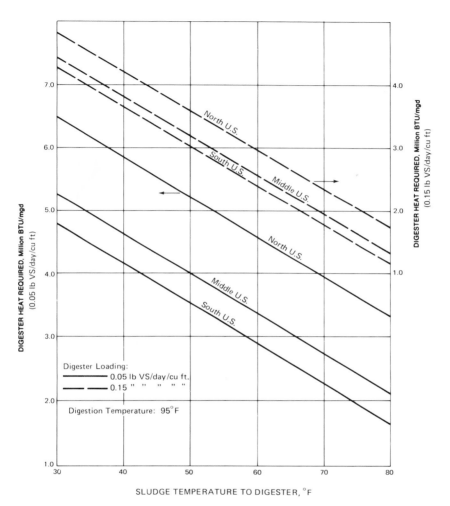

Figure 5-7. Anaerobic digester heat requirements for primary plus waste-activated sludge.

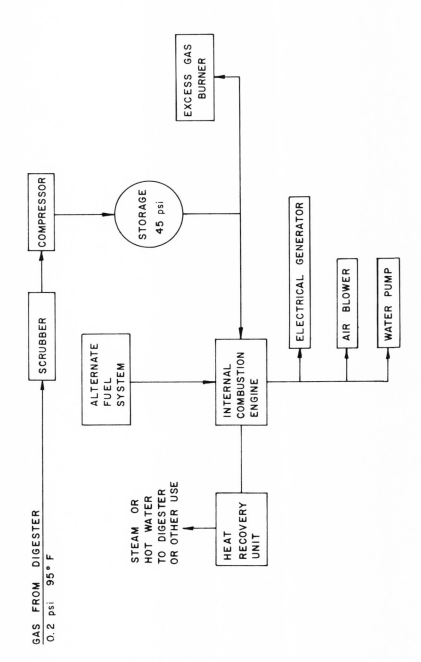

Figure 5–8. Anaerobic digester gas utilization system.

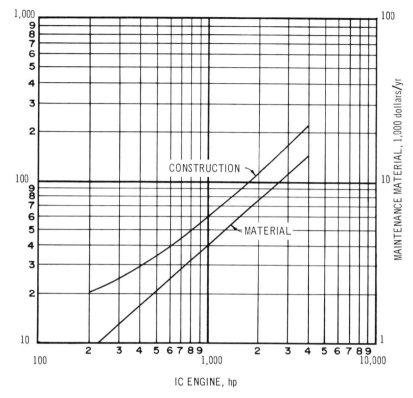

Figure 5-9. Construction and material costs for internal combustion engine (600-rpm engine with heat recovery and alternate fuel system).

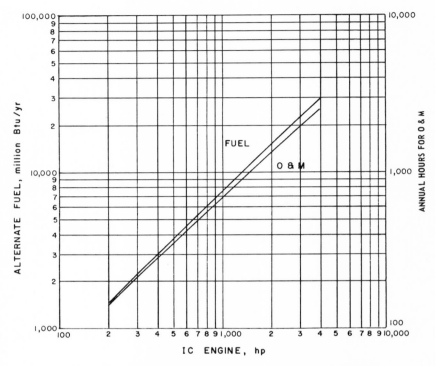

Figure 5-10. Labor and alternate fuel requirements for internal combustion engine (600-rpm engine with heat recovery and alternate fuel system).

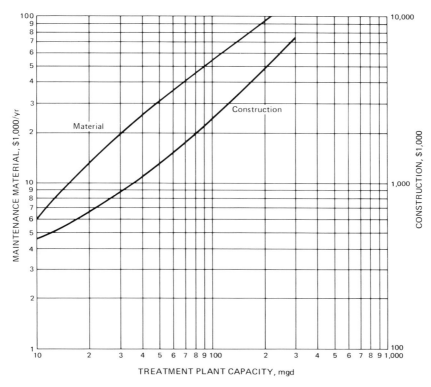

Figure 5–11. Construction and material costs for digester gas utilization system (complete system for electricity generation as shown in Figure 5–8).

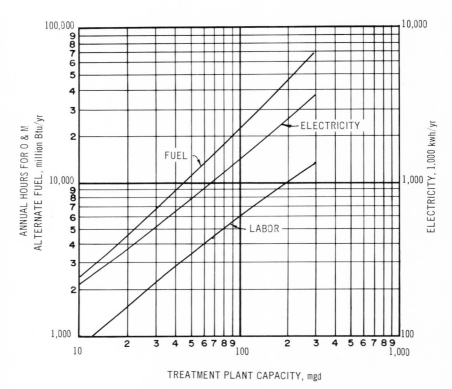

Figure 5-12. Labor and energy requirements for digester gas utilization system (complete system for electrical generation as shown in Figure 5-8).

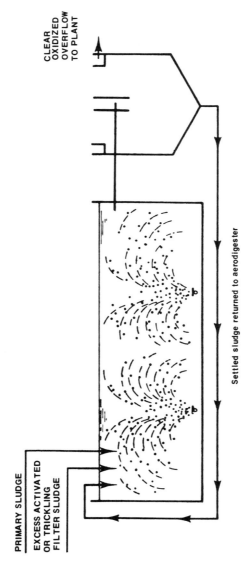

Figure 5-13. Aerobic digestion system (reference 6).

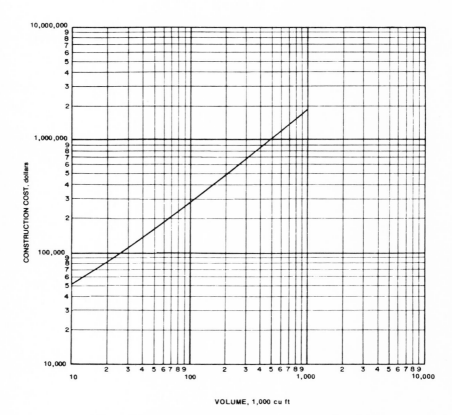

Figure 5-14. Construction costs for aerobic digestion.

Sludge Stabilization/141

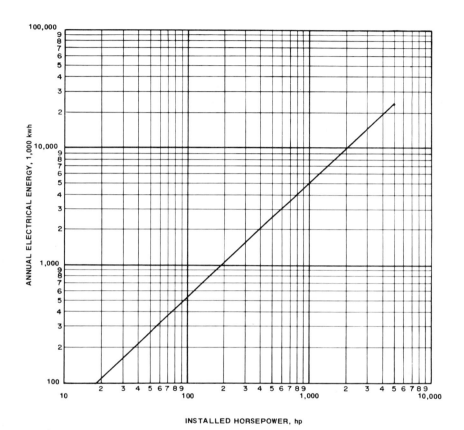

Figure 5-15. Energy requirements for aeration basin and mechanical aeration equipment.

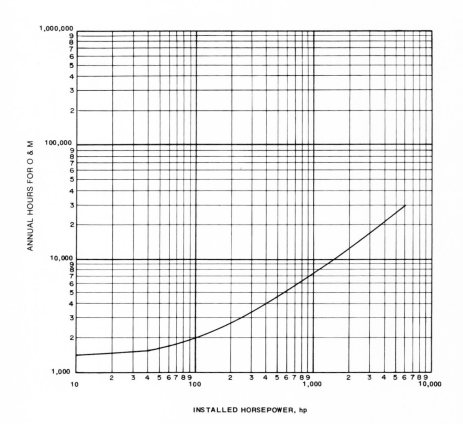

Figure 5-16. Labor requirements for aeration basin and mechanical aeration equipment.

Sludge Stabilization/143

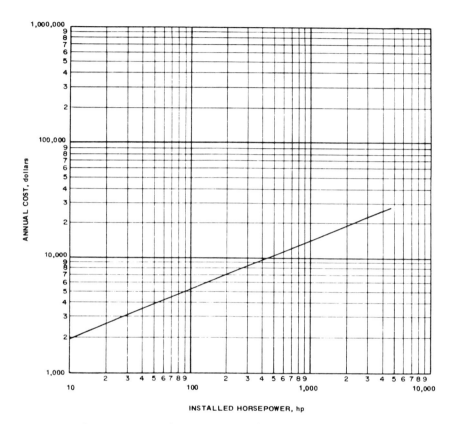

Figure 5-17. Maintenance material and supply costs for aeration basins and mechanical aeration equipment (Sept. 1976).

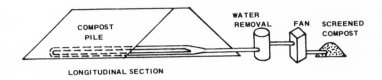
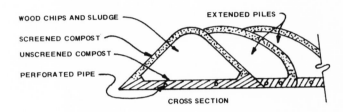

Figure 5-18. Static pile composting as developed by the Agricultural Research Service at Beltsville, Maryland.

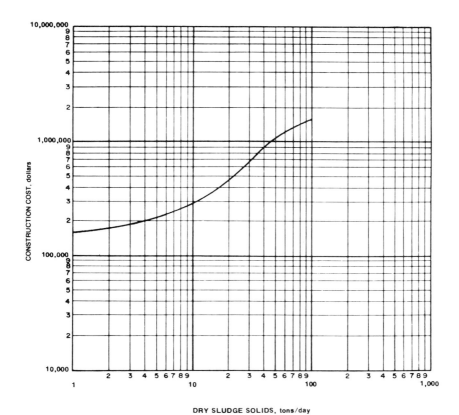

Figure 5-19. Construction costs for sludge composting (Sept. 1976).

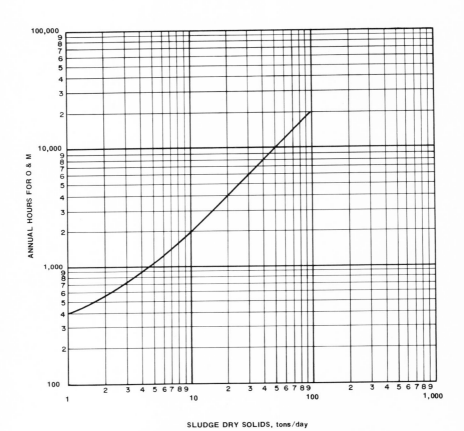

Figure 5-20. Labor requirements for sludge composting.

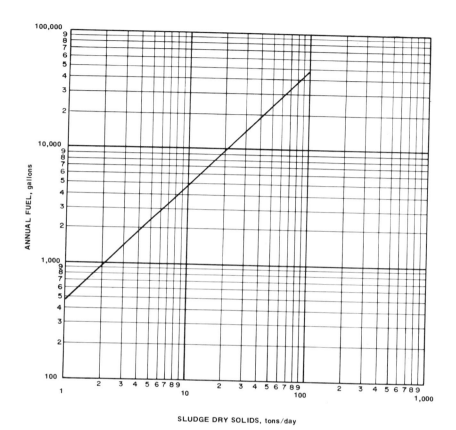

Figure 5-21. Fuel requirements for sludge composting.

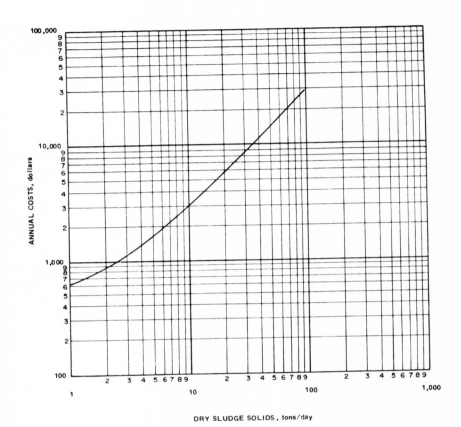

Figure 5-22. Maintenance material and supply costs for sludge composting (Sept. 1976).

Chapter 6

Sludge Incineration and Drying

INCINERATION

An incinerator is usually part of a sludge-treatment system which includes sludge thickening, a macerating or disintegrating system, a dewatering device (such as a vacuum filter, centrifuge, or filter press), an incinerator feed system, air-pollution-control devices, ash-handling facilities, and the related automatic controls. Important considerations in evaluating incineration costs include the composition of the sludge feed and the amount of auxiliary fuel required. Air-pollution constraints and resultant equipment and treatment requirements, as well as ash disposal, are also important.

Of major interest from the standpoint of sludge incineration is the heat value of the sludge, which is summarized in Table 6-1. The combustible portion of sewage sludge has a BTU content approximating that of lignite coal. When a sludge with a moisture content of about 75% is delivered to the incinerators (3 pounds of water for each pound of dry solids), the heat required to evaporate the water nearly balances the available heat from combustion of the dry solids.

The two major incineration systems employed in the U.S. are the multiple-hearth furnace and the fluidized-bed incinerator. The drying and combustion processes which occur in the furnace consist of the following phases: (a) raising the temperature of the feed sludge to 212°F, (b) evaporating water from the sludge, (c) increasing the water vapor and air temperature of the gas, and (d) increasing the temperature of the dried sludge volatiles to the ignition point. Practical operation of an incinerator requires that air in excess of theoretical requirements be supplied for complete combustion of the fuel. The introduction of excess air has the effect of reducing the burning temperature and increasing the heat losses from the furnace.

Heat is emitted by the burning of sludge in a furnace. Some of this heat is absorbed by the furnace and lost by radiation. A large portion of the emitted heat is lost with the stack gases, while a small portion is lost

TABLE 6-1
High Heat of Combustion of Sludges[a]

Material	Combustibles (%)	Ash (%)	Average BTU/lb
Grease and scum	88.5	11.5	16,750
Raw sewage solids	74.0	26.0	10,285
Fine screenings	86.4	13.6	8,990
Ground garbage	84.8	15.2	8,245
Digested sewage solids and ground garbage	49.6	50.4	8,020
Digested sludge	59.6	40.4	5,290
Grit	33.2	69.8	4,000

[a] Computed on a total dry-solids, moisture-free basis.

with the ash. The heat lost in the stack gases is available for recovery and reuse for purposes such as heating the incoming sludge and air.

There are a number of variables which influence the amount of fuel required and the resulting cost of sludge incineration. The principal variables are the moisture and volatile-solids content of the sludge. Their effect on the amount of fuel required for incineration is shown by Figure 6–1.* Temperatures of 1,350°–1,400°F are generally accepted as necessary to insure deodorization of the stack gases of a conventional incinerator. To insure complete thermal oxidation, it has been found necessary to maintain 50%–100% excess air over the stoichiometric amount of air required in the combustion zone. This excess air is undesirable because it pirates 12%–24% of the input BTUs for heating the excess air. If excess air is not supplied, it is difficult to maintain the minimum deodorizing temperature. Therefore, a closely controlled minimum excess-air flow is desirable for maximum thermal economy. The amount of excess air required varies with the type of incineration equipment, the nature of the sludge to be incinerated, and the disposition of the stack gases. The impact of the use of excess air on the fuel required for sludge incineration is shown in Figure 6–2.

The stack gases leaving the incinerator represent a potential source of energy. By passing the gases through a heat exchanger it is possible to

---

*Illustrations for this chapter start on page 164.

extract heat for use in preheating the incoming furnace air, in sludge conditioning by heat treatment, or for other uses in the plant. Electricity can be generated by use of a boiler-generator system fueled by the stack-gas heat. Figure 6-3 presents the potential for net recovery of heat by heat-exchange equipment installed on a sludge incinerator. This analysis of heat recovered is independent of the type of incinerator used for combustion of sludge because only the combustion products or flue gases are considered. The potential for energy recovery from stack gases is significant. A detailed analysis of a 30-million-gallons-per-day activated-sludge plant showed that all the electrical needs of the plant (about 8,300,000 kilowatt-hours per year) could be met by generating electricity from the incinerator stack gases (1,400° F initial flue gas temperature) from incineration of a 16% solids primary + WAS sludge.[39] Whether or not this would be a cost-effective source of electrical energy would depend on local conditions.

A heat-treatment incineration system has been proposed (and is being installed in three U.S. plants) which eliminates the need for any auxiliary fuels.[45] The heat-treatment conditioning enables an autogenous sludge cake to be achieved. As can be seen from Figure 6-4, heat production exceeds that required for combustion as typical primary + WAS concentrations exceed 25% solids (69% volatile). Solids concentrations of 30%-40% are often achieved following heat treatment. Thus, there may be sufficient heat available in the stack gases to provide the heat needed for heat treatment, which results in a self-sustaining sludge-incineration system.

One approach to supplying the supplementary fuel needed for sludge incineration that has been suggested is to use solid wastes as fuel. The amount of solid waste required to sustain combustion of sludges is shown in Figure 6-5 based on 25% moisture in the solid waste and 4,750 BTUs per pound of solid waste. Sludge with 5% solids and 70% volatile solids would require 28% refuse to sustain combustion.

The Kansas City metropolitan area is considering incineration of 750-1,000 tons per day of shredded, air-classified refuse and dried sludge (85% solids) for generation of electricity. Ferrous metals and possibly aluminum would be recovered. Use of a suspension-fired water-wall incinerator and sale of the electricity provides a potential economic savings of about 28% over separate refuse disposal in a landfill.

Air-pollution concerns must be addressed when considering incineration. Detailed data have been presented for several municipal sludge installations.[6,47] The major categories of concern are particulates, metals, gaseous pollutants, and organic compounds.

National air-pollution standards for discharges from municipal sludge incinerators have been promulgated which limit emissions of particulates (including visible emissions) from incinerators used to burn wastewater sludge as follows:

1. No more than 0.65 grams per kilogram of dry sludge input (1.30 pounds per ton of dry sludge input).
2. Less than 20% opacity.[48]

Available data indicate that, on the average, uncontrolled multiple-hearth incinerator gases contain about 0.6 grain of particulate per standard cubic foot of dry gas.[49] Uncontrolled fluid-bed reactor gases contain about 1.0 grain of particulate per standard cubic foot.[14] For average municipal wastewater sludge, this corresponds to about 33 pounds of particulates per ton of sludge burned in a multiple hearth, and about 45 pounds of particulates per ton of sludge burned in a fluid-bed incinerator. Particulate collection efficiencies of 96%–97% are required to meet the standard, based on the above uncontrolled-emission rate. Venturi scrubbers have the demonstrated capability to meet the particulate discharge requirement without a significant increase in electrical power requirements.[50]

Most metals present in municipal sludges are converted to oxides which appear in the particulates removed by the scrubber or in the ash. Lead and mercury are the only two metals which vaporize to an extent that the stack-gas concentrations would be increased. However, it has been found that less than 15% of the lead and 2% of the mercury appear in the flue gas at the Palo Alto, California, sludge incinerator.[45] The Environmental Protection Agency has set a standard of 3,200 grams of mercury per day for discharge from a sewage-sludge incinerator. The Palo Alto incinerator discharge was only 6 grams per day. The per capita lead discharge was equivalent to the lead discharged from driving an auto using unleaded gas a distance of 200 feet per day. Metal discharges should not present a limitation as properly designed and operated municipal systems have met all air-pollution standards for metals.

Gaseous pollutants which could be released by sludge incineration are hydrogen chloride, sulfur dioxide, oxides of nitrogen, and carbon monoxide. Carbon monoxide is no threat if the incinerator is properly designed and operated. Hydrogen chloride, generated by decomposition of certain plastics, is not a significant problem at concentrations currently observed. Consideration of the possibility of SO_2 and NO_x pollution is aided by examination of the sulfur and nitrogen content of sludges. Sulfur content is relatively low in most sludges. In addition, much of this sulfur is in the form of sulfate, which originated in the wastewater. Sulfur dioxide is not expected to be a serious problem. Sludge typically has a high nitrogen content from proteinaceous compounds and ammonium ions. Limited data are available for predicting whether a high proportion of these materials will be converted to oxides of nitrogen by sludge incineration. Concentrations should be less than 100 ppm from a properly operated incinerator and were observed to be less than 10 ppm from one facility tested by the EPA.[48] Considering this low concentration, the production of oxides of nitrogen will probably not limit the use of incineration for disposing of sludge in most cases. The amount of NO_x per capita generated by a sludge incinerator has been equated to that generated by driving an auto less than 0.1 mile under the 1975 Federal NO_x standards.[45]

Toxic substances, such as pesticides and PCBs, could be discharged from the organic substances in the sludge. Tests have shown, however, that total destruction of PCBs was possible when oxidized in combination with sewage sludge and with exhaust-gas temperatures of 1,100°F. A 95% destruction of PCBs was achieved in a multiple-hearth furnace with no afterburning at exhaust temperatures of 700°F.

The EPA Sewage Sludge Incineration Task Force concluded that it has been adequately demonstrated that existing well-designed and operated municipal wastewater-sludge incinerators equipped with an adequate scrubbing system are capable of meeting the most stringent particulate-emission-control regulation existing in any state or local control agency.[51] This observation, coupled with the fact that the newly promulgated federal standards are based on demonstrated performance of an operating facility, indicates that use of proper emission controls and proper operation of the incineration system will enable a facility to meet all existing air-pollution regulations.

The volume reduction by sludge incineration is over 90% when compared to the volume of dewatered sludge. The ash from the incineration process is free of pesticides, viruses, and pathogens. The metals in the ash are approximately at the same ratio as in the raw sludge; however, the metals are now in the less soluble oxide form. The ash can readily be transported in the dry state to appropriate landfill sites.

Multiple-Hearth Incineration

The multiple-hearth furnace is the most widely used wastewater-sludge incinerator in the U.S. today because it is simple and durable, and has the flexibility of burning a wide variety of materials even with fluctuations in the feed rate. A typical hearth furnace is shown in Figure 6–6 and consists of a circular steel shell surrounding a number of solid refractory hearths and a central rotating shaft to which rabble arms are attached. The operating capacity of these furnaces is related to the total area of the enclosed hearths. They are designed with diameters ranging from 54 inches to 21 feet 6 inches, and containing 4–12 hearths. Table 6–2 summarizes the characteristics of standard multiple-hearth furnaces. Capabilities of multiple-hearth furnaces vary from 200 to 8,000 pounds per hour of dry sludge, with operating temperatures as high as 1,700° F. The dewatered sludge enters at the top through a flapgate and proceeds downward through the furnace from hearth to hearth through the rotary action of the rabble arms.

Capital costs for multiple-hearth sludge-incineration systems are shown in Table 6–3 and Figure 6–7. Equipment costs include the furnace, controls, scrubbing system, dry cyclone, and auxiliaries on an installed basis. Labor associated with incineration facilities provides for operation and maintenance of incinerators and accessory facilities, including sludge conveyors, ash-handling equipment, control center, and the enclosing structures. Figure 6–8 presents estimated manpower requirements. Figure 6–9 presents operating fuel requirements for different sludge types. Start-up fuel requirements are shown in Figure 6–10 and must be added to the operating fuel determined from Figure 6–9. Total costs for sludge incineration are typically $45–$80 per ton of dry solids.

Fluidized-Bed Incineration

The first fluidized-bed wastewater-sludge incinerator was installed in 1962, and there are now several units operating. They range in size from 220 to 5,000 pounds of dry solids per hour. A typical section of a fluid-bed reactor used for combustion of wastewater sludges is shown in Figure 6–13. The fluidized-bed incinerator is a vertical cylindrical vessel with a grid in the lower section to support a sandbed. Dewatered sludge is injected above the grid and combustion air flows upward at a pressure of 3.5–5.0 psig and fluidizes the mixture of hot sand and sludge. Supplemental fuel can be supplied by burners above or below the grid. In essence, the reactor is a single-chamber unit where both moisture evaporation and combustion occur at 1,400°–1,500° F in the sandbed. All the combustion gases pass through the 1,500° F combustion zone with residence times of several seconds. Ash is carried out the top with combustion exhaust and is removed by air-pollution-control devices.

The quantity of excess air is maintained at 20%–25% to minimize its effect on fuel costs. The heat reservoir provided by the sandbed enables

TABLE 6-2
Standard Sizes of Multiple-Hearth Furnace Units[a]

Effective hearth area (sq ft)	Outer diameter (ft)	No. hearths	Effective hearth area (sq ft)	Outer diameter (ft)	No. hearths
85	6.75	6	988	16.75	7
98	6.75	7	1041	14.25	11
112	6.75	8	1068	18.75	6
125	7.75	6	1117	16.75	8
126	6.75	9	1128	14.25	12
140	6.75	10	1249	18.75	7
145	7.75	7	1260	16.75	9
166	7.75	8	1268	20.25	6
187	7.75	9	1400	16.75	10
193	9.25	6	1410	18.75	8
208	7.75	10	1483	20.25	7
225	9.25	7	1540	16.75	11
256	9.25	8	1580	22.25	6
276	10.75	6	1591	18.75	9
288	9.25	9	1660	20.25	8
319	9.25	10	1675	16.75	12
323	10.75	7	1752	18.75	10
351	9.25	11	1849	22.25	7
364	10.75	8	1875	20.25	9
383	9.25	12	1933	18.75	11
411	10.75	9	2060	20.25	10
452	10.75	10	2084	22.25	8
510	10.75	11	2090	18.75	12
560	10.75	12	2275	20.25	11
575	14.25	6	2350	22.25	9
672	14.25	7	2464	20.25	12
760	14.25	8	2600	22.25	10
845	16.75	6	2860	22.25	11
857	14.25	9	3120	22.25	12
944	14.25	10			

[a] No. of Hearths (NHEAR): 6 to 12.
Wall Thickness: 13.5 inches.
Outer Diameter (HDIA): 6.75–22.25 feet.
Effective Hearth Area (FHA): 85–3120 square feet.

TABLE 6-3
Estimated Construction Costs of Multiple-Hearth Incinerators (Sept. 1976)

Cost component	Total effective hearth area (sq ft)				
	180	440	1050	1980	4100
Manufactured equipment	$675,000	$950,000	$1,558,000	$1,920,000	$2,600,000
Labor	230,000	330,000	545,000	672,000	910,000
Housing	30,000	50,000	85,000	120,000	175,000
Electrical and instrumentation	140,000	200,000	328,000	407,000	553,000
Miscellaneous items	161,000	230,000	377,000	468,000	636,000
Contingency	185,000	264,000	434,000	538,000	731,000
Total Estimated Cost	$1,421,000	$2,024,000	$3,327,000	$4,125,000	$5,605,000

reduced start-up times when the unit is shut down for relatively short periods (overnight). As an example, a unit can be operated 4–8 hours a day with little reheating when restarting, because the sandbed serves as a heat reservoir.

Exhaust gases are usually scrubbed with treatment plant effluent, and ash solids are separated from the liquid in a hydrocyclone, with the liquid stream returned to the head of the plant.

Actual field cost data on fluidized-bed systems for municipal sludges are limited. They are often competitive with multiple-hearth systems on capital costs. Figures 6–14 and 6–15 present data on fuel and energy consumption.[39]

Wet-Air Oxidation

The heat-treatment, sludge-conditioning system illustrated in Figure 2–13 can be used for sludge reduction of oxidation by operation at higher temperatures (350°–400°F) and higher pressures (1,200 psig). The wet-air oxidation (WAO) process is based on the fact that any substance capable of burning can be oxidized in the presence of liquid water at temperatures between 250°F and 700°F. Wet-air oxidation does not require preliminary dewatering or drying as in conventional combustion processes. However, the oxidized ash must be separated from the water by vacuum filtration, centrifugation, or some other solids-separation technique. Air pollution is minimized because the oxidation takes place in water at low temperatures and no flyash, dust, sulfur dioxide, or nitrogen oxides are formed.

The problems noted earlier in the heat-treatment discussion—recycle of high-strength liquors to the wastewater treatment plant, the presence of refractory materials, high maintenance costs, and odor control—also exist for the WAO application. The high-pressure/high-temperature system also introduces some significant safety concerns. The cost of the system for sludge reduction is usually higher than competitive reduction systems.[52] Use of heat treatment for sludge conditioning is more widespread than use of WAO as a reduction process.

Lime Recalcining

Lime is often used as a coagulant, either as a tertiary step or ahead of the primary clarifier in either a biological or a physical-chemical plant for removal of phosphorus from wastewaters. There is, of course, considerable experience around the world with the successful recalcining and reuse of lime used in water-treatment plants, and these techniques may also be used to recalcine and reuse lime in wastewater applications.

The process of recalcining consists of heating the dewatered calcium-containing sludge to about 1,850°F, which drives off water and carbon dioxide leaving only the calcium oxide (or quicklime). Either multiple-hearth or fluidized-bed furnaces may be used for recalcining. Recovery and reuse of the lime reduces the amount of chemical sludge requiring disposal by a factor of about 20. The significant savings in sludge-disposal costs may offset enough of the costs of lime recalcining to make the economics attractive. The economic feasibility of lime recalcining must be carefully evaluated for each locale, however. In cases where there are acceptable sludge landfill sites, it has proven to be cost effective to dewater the lime sludges and bury them rather than recover them by recalcining.

Figure 6–16 presents information on fuel and energy requirements for multiple-hearth recalcination. Figures 6–7 and 6–12 may be used for capital costs and maintenance material costs.

A more detailed discussion of lime recalcining systems and costs is available.[53]

Pyrolysis

There is a recent surge of interest in pyrolysis as a means of disposing of sewage sludge. This has come about as a result of the apparent need for new and improved processes and equipment in the practice of sludge disposal, and the possibility that pyrolysis may offer an alternative to incineration which may be lower in cost, use less fuel, provide improved air-pollution control, and afford greater heat recovery, under certain conditions.

Pyrolysis is a process in which organic material is decomposed at high temperature in an oxygen-deficient environment. The action, causing an irreversible chemical change, produces three types of products: gas, oil, and char (solid residue). Water vapor is also produced, usually in relatively large amounts depending on the initial moisture content of the materials being pyrolysed. Residence time, and temperature and pressure in the reactor, are controlled to produce various combinations and compositions of the products. Two general types of pyrolysis processes may be used. The first, true pyrolysis, involves applying all required heat externally to the reaction chamber. The other, sometimes called partial combustion and gasification, involves the addition of a small amount of air or oxygen directly into the reactor. The oxygen sustains combustion of a portion of the reactor contents, which in turn produces the heat required to dry and pyrolyse the remainder of the contents.

Pyrolysis of municipal refuse and of sewage sludge has been considered as a means for ultimate disposal of wastes for several years.[54-57] The results of various studies and pilot programs indicate that if the moisture content of a sludge is below 70%-75% enough heat can be generated by combustion of the oil and gases produced from the pyrolysis of sludge for the process to be thermally sustaining. Pyrolysis of municipal refuse, and combinations of refuse and wastewater sludges, will provide energy in excess of that required in the pyrolytic process.[55, 57]

Laboratory, pilot, and demonstration systems for pyrolysis of wastewater sludges have been tested but no data from full-scale systems are available. Therefore, the data presented must be considered preliminary. Pyrolysis systems are in the developmental stages and additional information will become available as research and development work and the operation of full-scale plants progresses.

Research and development work has been conducted using multiple-hearth furnaces, similar in design to conventional sludge incinerators, for pyrolysis of wastewater sludges mixed with municipal solid wastes. A 100-ton-per-day unit is being installed at Concord, California.[58] Shredded and classified solid wastes and dewatered sludge are fed to the furnace either in a mixture or separately, with the wetter sludge fed higher in the furnace. Recirculated hot shaft cooling air and supplemental outside combustion air are fed to the lower hearths to sustain partial combustion of the wastes circulating down through the furnace. Fuel gas produced through the pyrolysis reaction is then burned in a high-temperature afterburner. The resulting heat can be used in a waste-heat boiler to produce high-pressure steam. It may also be possible to burn the fuel gases directly in a boiler. Char from the process is not used, but because it has some fuel value, it may be usable as an industrial fuel.

The multiple-hearth process offers the following advantages: (1) it is usable in much smaller plants than most other pyrolysis systems; (2) it employs modifications of well-developed sludge incineration equipment; (3) it produces high-temperature gases without raising temperatures in the solid phase to the slagging point; and (4) the conversion from existing conventional sludge-incineration systems is a relatively simple procedure. Its disadvantages include: (1) the fuel value of the char is not used; (2) the high-temperature fuel gases must be used on site; and (3) the incoming solid wastes must be well classified.

Estimates of the potential energy production from pyrolysis of refuse and sludge combined have been made for representative systems. Process differences result in variations in the composition and quantities of fuel produced, but should result in relatively minor variations in net heat output. The estimates made indicate that the pyrolysis process would be self-sustaining from an energy standpoint with mixtures of sludge and solid wastes containing 25%–40% sludge and 60%–75% solid wastes. The refuse-to-sludge ratio for a typical residential community is in the range of 10:1 to 15:1 on a dry-solids basis, and 3:1 to 8:1 on a wet-solids basis, indicating that more than enough refuse is generally available for mixing with sludge to operate the process without the need for an external energy source.

Pyrolysis appears to have several advantages over incineration. For example, some pyrolysis processes can convert wastes to storable, transportable fuels such as fuel gas or oil, while incineration only produces heat that must be converted to steam. Pyrolysis gives a 50% greater reduction in volume of residue over incineration, and the residue is a more readily usable by-product. Air pollution is not as severe a problem in pyrolysis systems because the volume of stack gases and the quantity of particulates in the stack gases are less.

On the other hand, pyrolysis is essentially still in the developmental stage. Most of the pyrolytic fuel gases have relatively low heat values and the pyrolytic oil is corrosive, requiring that it be mixed with other fuel oil for best results.

The construction and operating costs for most pyrolysis systems are much more uncertain than for incineration. Reliable cost data for pyrolysis systems will not be available until significant operating experience is developed from the ongoing and planned demonstration projects.

DRYING OF SLUDGE

Flash Drying

Flash drying is the instantaneous removal of moisture from solids by introducing them into a hot gas stream. This process was first applied to the drying of wastewater sludge at the Chicago Sanitary District in 1932. A flow diagram of the process is shown in Figure 6–17.[59] Originally, units were designed to dry sludge for fertilizer and burn only the excess. The system is based on three distinct cycles, which can be combined in different arrangements. The first cycle is the flash-drying cycle, where wet filter cake is blended with some previously dried sludge in a mixer to improve pneumatic conveyance. The blended sludge and the hot gases from the furnace at 1,300°F are mixed ahead of the cage mill and flashing of the water vapor begins. The cage mill mechanically agitates the mixture of sludge and gas and the drying is virtually complete by the time the sludge leaves the cage mill. At this stage, the sludge is at a moisture content of 8%–10%, and dry sludge is separated from the spent drying gases in a cyclone. The dried sludge can be sent either to fertilizer storage or to the furnace for incineration.

The second cycle is the incineration cycle. Combustion of fuel is essential to provide heat for drying the sludge, and the fuel may be gas, oil, coal, or wastewater sludge. Primary combustion air, provided by the combustion air fan, is preheated and introduced at a high velocity to promote complete sludge combustion.

The third cycle is the effluent gas cycle, or induced-draft cycle, consisting of the deodorizing and combustion air preheaters, dust collector, induced-draft fan, and stack. Heat recovery is practiced to improve economy. The effluent gases then pass through a dust collector (dry centrifuge or wet scrubber) and the inducing fan discharges the effluent gases through a stack into the atmosphere.

Perhaps the most notable current U.S. usage of this process is that by the city of Houston, Texas, primarily for drying sludge for use as a fertilizer.[59] After complete processing the dry product has a moisture content of around 5.5%. From analysis at the time of sales of fertilizer in January 1972, the moisture content was 5.0%; ash, 34.76%; nitrogen, 5.34%; and available phosphoric acid, 3.93%. The ash content fluctuates;

the lowest on record is 26.4% and the highest, 44.3%. Throughout the experience with this operation, the city's marketing arrangements have been scheduled on the basis of competitive bidding. The successful bidder is committed to placing orders with the city for its entire production for the contract period of 5 years. The material is shipped in bulk by railroad car lots or sometimes by barge. It is bagged for resale at the point of arrival. The present contractor has been handling it for about 10 years, disposing of about 80% of the production in the citrus groves of Florida. There has never been a time when it was not possible to dispose of the entire sludge production by sales.

The use of the flash-drying systems for incineration alone has not proven attractive. The Metropolitan Denver Sewage Disposal District No. 1 plant (approximately 100 million gallons per day in capacity) abandoned a system of this type because of air pollution and problems of continuing explosions in the units. As an incineration system, flash drying has the disadvantages of complexity, potential for explosions, and potential for air pollution by fine particles. An advantage is the flexibility it offers for drying a portion of the sludge for fertilizer.

Flash drying is relatively expensive because of fuel costs (contrasted to incineration—no heating value is realized from the sludge) and because pretreatment needs for production of sludge, which must have some reasonable nutrient balance, are also expensive. Fuel consumption for production of dried sludge is about 8,000 BTUs per pound for flash drying.

Many flash-drying installations have been abandoned because of their high costs (typically twice or more the cost for incineration and heat recovery), odor problems, and the problems associated with the fine particulates (air pollution and explosions).

Another approach to heat drying of wastewater sludges for use as fertilizers has been studied at the Blue Plains plant in Washington, D.C.[60] A schematic of the system is shown in Figure 6-18. Drying is achieved in a jet mill in this case. The mill has no moving parts and offers the ability to dry and classify solids simultaneously.

Sludge is dried and sterilized at a temperature of 1,100°F. In order to obtain a fertilizer with the desired nutrient balance at the Blue Plains plant, it has been necessary to supplement the nitrogen content of the sludge. The resulting product contained 6% nitrogen, 4% phosphoric acid, and no potash.

During the system's break-in phase, more than 15,000 tons of sludge were processed, at a running rate of 200–270 wet tons of sludge per day, with production of dry product per day being 40–54 tons.[60] The solids

162/Handbook of Sludge-Handling Processes

content of the feed sludge averaged 22%. Operation and maintenance problems have resulted in the temporary shutdown of this unit at the Blue Plains plant. Trash and fibrous material from the primary clarifiers have caused problems of fires and materials handling. Very serious erosion of the drying unit has been encountered. In addition, the product has been extremely dusty, thereby limiting its marketability.

Capital cost data for the jet-mill system are unavailable at this time. The cost of operating the unit at Blue Plains has been as shown in Table 6–4, adjusting fuel oil costs to $0.50 per gallon.[60]

Unit costs for power, labor, and nitrogen were not reported,[60] so these cost items may not be on a basis comparable to costs presented for other processes in this report.

Preliminary data from the Blue Plains plant indicate that with a feed sludge at 78% water, approximately 60 gallons of No. 2 fuel oil (8,500,000 BTUs) are required per ton of dry sludge processed (13 gallons per wet ton processed).

Solvent Extraction

A system (Basic Extractive Sludge Treatment, or BEST) for drying sludge to 95% solids is under development by Resources Conservation Company.[60, 61-63]

In addition to drying the sludge, greases and oils are recovered for possible use as an energy source or commercial by-product. Sludge is introduced into the BEST system, stabilized at about 50°F, and mixed with cold recycled solvent. The system uses triethylamine ("TEA") as the solvent in the primary dewatering step (see Figure 6–19). The solvent and water are completely miscible at temperatures below 65°F and are im-

TABLE 6–4
Blue Plains Operating Costs

Cost component	Operating costs only (per dry ton)
No. 2 fuel oil	$30.00
Electricity	9.08
Labor	6.81
Nitrogen supplement	10.67
Total	$56.56

miscible above that point. When the solvent and sludge are mixed below 65°F, the solids are easily separated by either centrifuge or a filter. Because the solvent-sludge mixture is warmed to about 60°F by the addition of heat of solution, it is chilled to 50°F before being fed to a solid-bowl centrifuge. After the centrifuge separates the solvent-water-solids mixture, solids go to a dryer and the liquid fraction to a decanter. The closed-cycle dryer removes the solvent and produces dry solids (95% solids, sterile); the solvent driven off is condensed and returned to the system. The water-solvent fraction is heated and moved to a decanter where the solvent and water form two layers. This solvent is recycled to the system, and the water layer is fed to a steam-stripping distillation column where the remaining fraction of solvent is recovered and returned to the system. The sterile dry solids produced by this process can be used as a fertilizer, a soil conditioner, or as a raw material, depending on the composition of the original sludge.

Heat and electricity requirements are a function of the solids fed and may be summarized as in Table 6–5.

For a typical sludge containing 16% oil, about 2,000 BTUs will be recovered in the form of oil per pound of dry solids. The recycle stream from a primary-WAS mixture is expected to have the following quality:

BOD = 2,760 milligrams per liter
COD = 7,335 milligrams per liter
SS = 400 milligrams per liter

The estimated costs for the BEST process for a primary-WAS mixture centrifuged to 100% solids are shown in Table 6–6 (based on data from the manufacturer gathered in pilot scale tests and yet to be confirmed on plant scale). If a potential value of $40 per ton for the recovered fertilizer could be achieved, it is apparent that the costs would be very attractive.

TABLE 6–5
Heat and Electricity Requirements of the BEST System as a Function of Solids Fed

% Solids in feed	Solids feed (kwh/lb)	Solids feed (BTU/lb)
7	0.52	3,514
11	0.38	3,266
20	0.26	3,088

TABLE 6-6
Estimated Costs of Drying by Solvent Extraction (BEST Process) for Primary + WAS at 11% Solids

Cost component	10 Ton day	25 Ton day	50 Ton day	100 Ton day
Amortized capital (X 0.0944)	$38.80	$28.97	$22.76	$18.88
Electric ($0.025/kwh)	19.00	19.00	19.00	19.00
Fuel ($3.00/MBTU)	19.60	19.60	19.60	19.60
Maintenance	12.33	9.21	7.23	6.00
Operations	8.63	4.60	3.45	2.30
Solvent (TEA at $0.89/lb)	2.78	2.78	2.78	2.78
Lime	2.10	2.10	2.10	2.10
Recycle stream treatment	4.06	3.68	3.44	3.34
Subtotal	$107.30	$89.94	$80.36	$74.00
Fuel credit (recovered oil)	(12.30)	(12.30)	(12.30)	(12.30)
Total	$95.00	$77.64	$68.06	$61.70

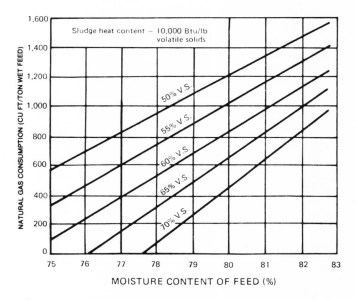

Figure 6-1. The effects of sludge moisture and volatile solids content on gas consumption (reference 6).

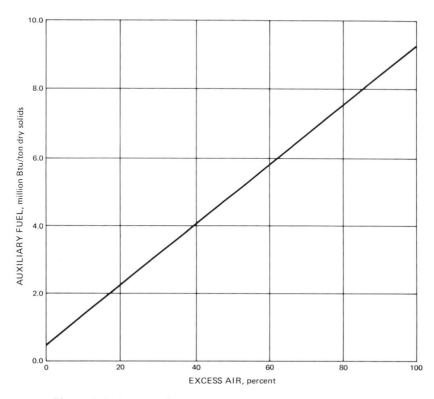

Figure 6-2. Impact of excess air on the amount of auxiliary fuel for sludge incineration. Assumptions: solids, 30%; exhaust temperature, 1,500°F; volatiles, 70% (reference 39).

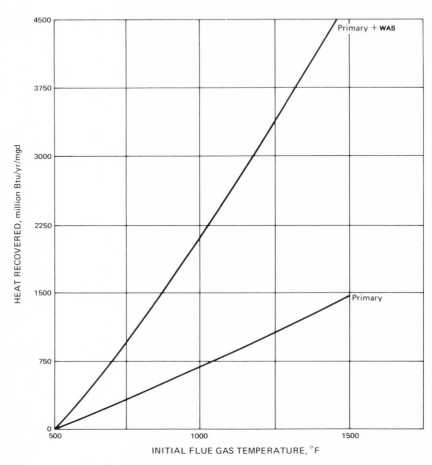

Figure 6-3. Potential heat recovery from incineration of sludge. Assumptions: final stack temperature = 500°F; 50% excess air (to convert BTU to kwh: 1 kwh = 10,500 BTU) (reference 39).

Sludge Incineration and Drying/167

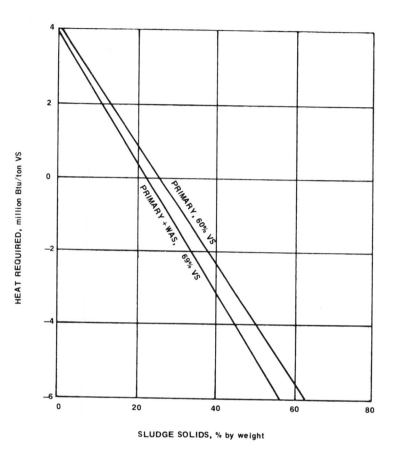

Figure 6-4. Heat required to sustain combustion of sludge (reference 39).

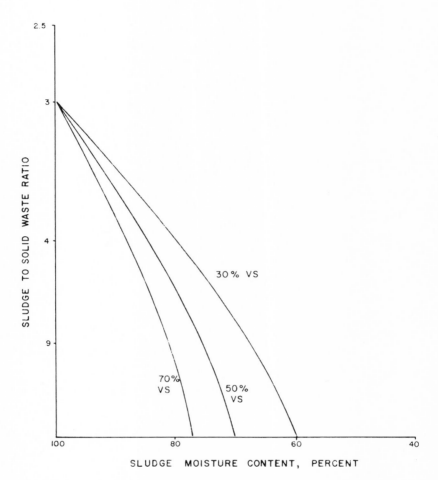

Figure 6-5. Combustion of sludge and solid waste. Assumptions: heat value of sludge, 10,000 BTU/lb VS; heat value of solid waste, 4,750 BTU/lb; moisture in solid waste, 25%; heat required to evaporate water in furnace, 2,100 BTU/lb water (reference 39).

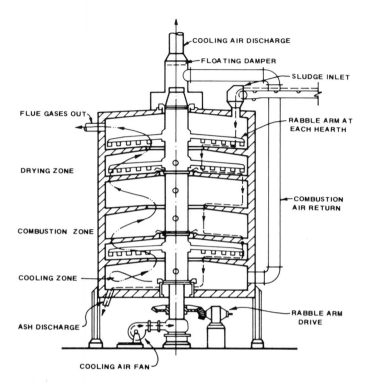

Figure 6-6. Cross section of a typical multiple-hearth incinerator (reference 6).

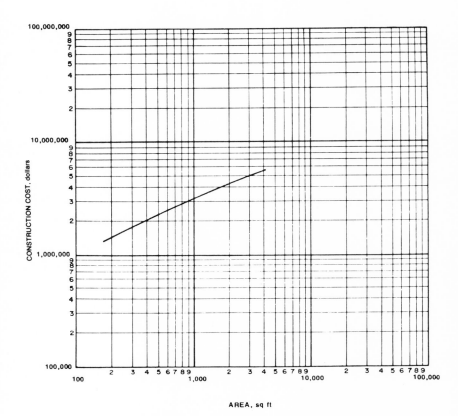

Figure 6–7. Construction costs for multiple-hearth incinerators (Sept. 1976).

Sludge Incineration and Drying/171

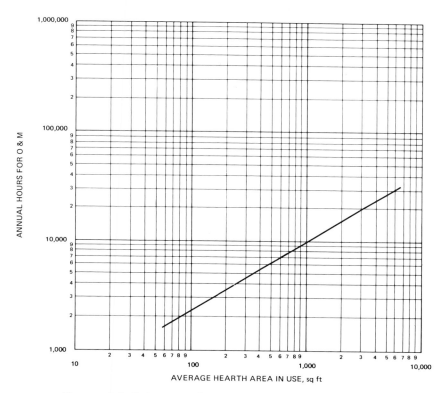

Figure 6-8. Labor requirements for multiple-hearth incineration (based on continuous operation).

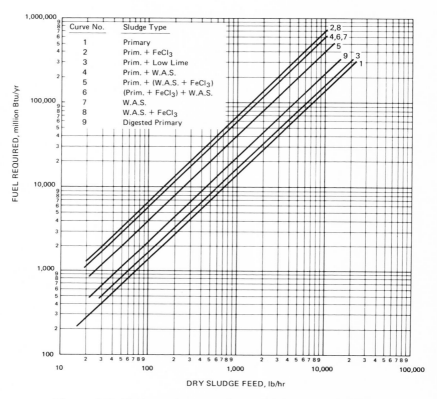

Figure 6-9. Fuel requirements for multiple-hearth furnace incineration (see figure 6-10 for start-up fuel). Operating parameters: incoming sludge temperature is 57°F; combustion temperature is 1,400°F; down time for cool-down equals start-up time; frequency of start-ups is a function of individual systems.

Sludge Incineration and Drying/173

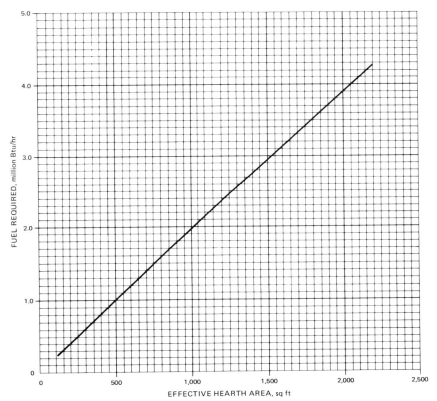

Figure 6-10. Multiple-hearth furnace incineration start-up fuel. Operating parameter: system operates 100% of the time. Design assumptions are as follows:

Solids concentration (%)	Loading rates (lb/hr/sq ft) (wet sludge)	
	Small plants	Large plants
14–17	6.0	10.0
18–22	6.5	11.0
23–30	7.0	12.0
31	8.0	12.0

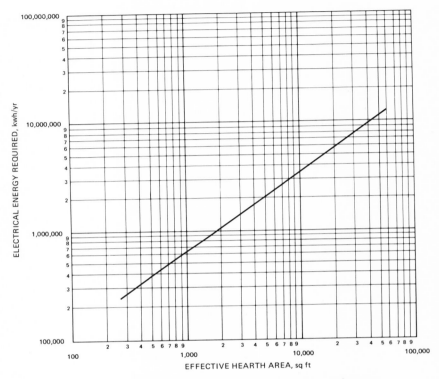

Figure 6–11. Energy requirements for multiple-hearth furnace incineration. Operating assumptions: heat-up time to reach 1,400° F; frequency of start-up is a function of individual system. Design assumptions: use in conjunction with Figure 6–9 to determine total fuel required; heat-up time is as follows:

Effective hearth area (sq ft)	Heat-up time (hr)
< 400	18
400–800	27
800–1400	36
1400–2000	54
> 2000	108

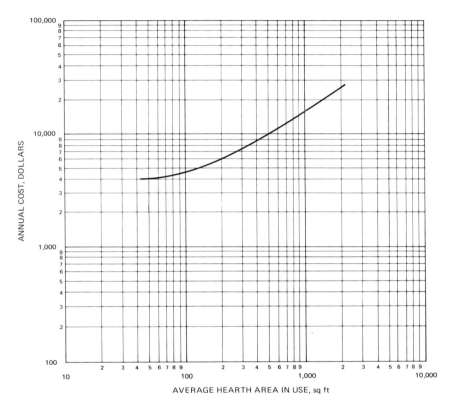

Figure 6–12. Maintenance material costs for multiple-hearth incineration (Sept. 1976).

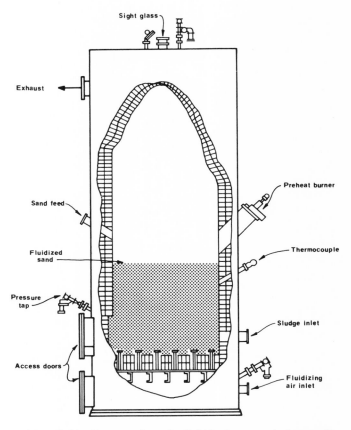

Figure 6-13. Cross section of a fluid-bed reactor (reference 6).

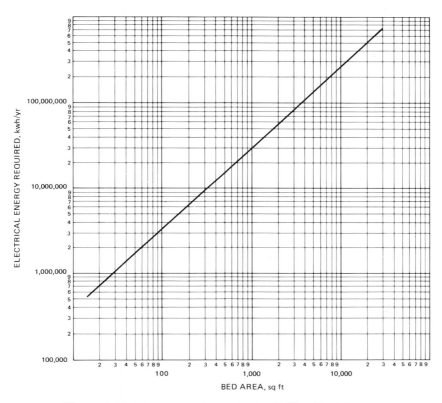

Figure 6–14. Energy requirements for fluidized-bed furnace incineration. Operating parameter: full-time operation.

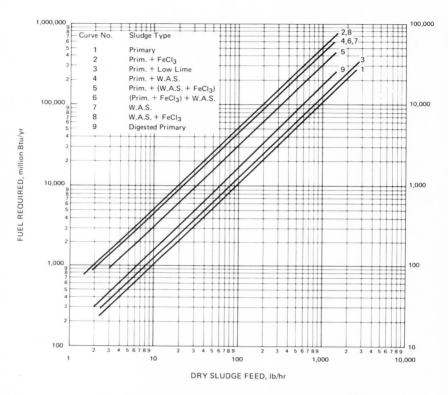

Figure 6–15. Fuel requirements for fluidized-bed furnace incineration. Operating conditions: combustion temperature is 1,400°F; 40% excess air, no preheater; start-up not included, 73,000 BTU/sq ft for start-up. Design assumptions: heat value of volatile solids is 10,000 BTU/lb; loading rates (lb/sq ft/hr):

Curve No.	Rate
1, 9	14
2, 4, 6, 7, 8	6.8
3	18
5	8.4

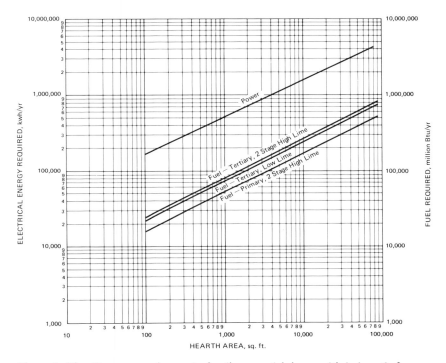

Figure 6-16. Energy requirements for lime-recalcining multiple-hearth furnace. Design assumptions: continuous operation; multiple-hearth furnace; 7 lbs/sq ft/hr loading rate (wet basis); gas outlet temperature = 900°F; product outlet temperature = 1400°F; power includes center shaft drive, shaft cooling fan, burner turbo-blowers, product cooler, and induced-draft fan.

Sludge composition	$CaCO_3$	$Mg(OH)_2$	Other inerts	Combustibles
Primary, 2-stage high lime	65%	2%	13%	20%
Tertiary, low lime	71	10	16	3
Tertiary, 2-stage high lime	86.1	4.3	6.1	3.5

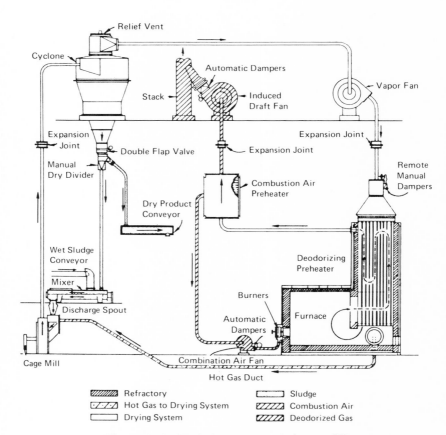

Figure 6-17. Flash-dryer system (reference 59).

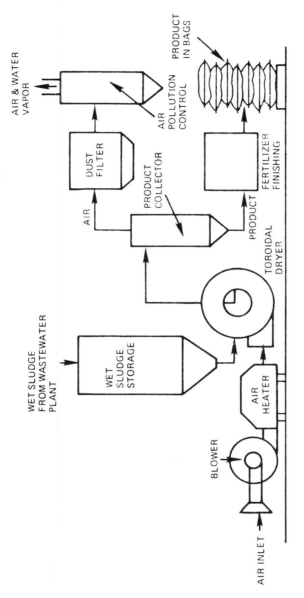

Figure 6–18. Sludge-drying system using the jet mill principle (reference 60).

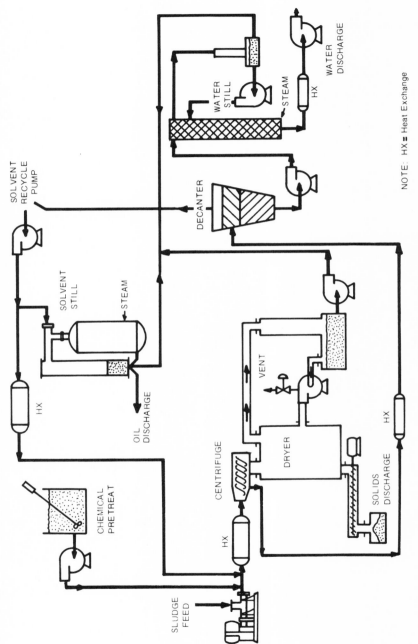

Figure 6-19. BEST System Schematic.

Chapter 7
Disposal and Land Application

The three major alternative modes for application of sludge to the land and their constraints have been defined as shown in Table 7-1.[6]

A survey of 176 landfill operations in 1972 found that only 30% permitted disposal of sewage sludge.[64] Despite this rather low percentage, stabilized sludge in landfills is recognized as an acceptable method of disposal.[6] The EPA estimates that 40% of wastewater sludges are currently disposed of in landfills and dumps, and it was projected that this percentage will be maintained in 1985.[1]

The landfill site's geology, hydrology, and soil conditions should be considered relative to the need for adequate protection of groundwater, conformation of area land-use planning, and provision of an adequate quantity of earth cover.[6]

Although past practice has emphasized disposal of ash or dewatered (i.e., 20% solids) sludge, a recent study indicates that liquid digested sludge (4% solids) can successfully be disposed of in a landfill.[65] With use of proper sludge-spreading techniques, the municipal solid waste had sufficient absorptive capability to retain the associated sludge moisture and prevent leachate generation. The entire liquid sludge production from Oceanside, California, has been disposed of in a landfill since 1972.

The costs for landfill of sludge are reported to be $1–$5 per ton at the landfill site.[6, 64] Transportation costs are discussed separately in the next chapter. Total costs at Oceanside for truck transport, unloading, and landfill disposal of liquid (4% solids) sewage sludge was $25–$32 per ton, economically competitive with other alternatives there.

Of course, landfilling is a disposal technique which makes no use of the nutrients in the sludge. The following section discusses land application which does reuse these nutrients.

It is not within the scope of this handbook to address the factors involved in land application of sludges and to review past experiences. Literally hundreds of references are available on the topic, many of which are listed in references 66 and 67. Current practices and much information

TABLE 7-1
Alternative Modes of Sludge Application to the Land, and Their Constraints

Disposal	Principal sludge form	Main constraints
Sanitary landfill	Dewatered cake or ash	Gas, leachate, and runoff control; land availability
Sites dedicated to sludge disposal	Liquid or dewatered	Leachate, runoff control, land availability
Cropland application	Liquid, cake dried, or compost	Application rate, unsatisfactory sludge
Land reclamation	Liquid or dewatered	Application rate, unsatisfactory sludge, availability of land

is contained in two seminar proceedings.[68, 69] Design guidance is also presented in reference 8. Major cities such as Denver[70] and Chicago[71] are applying their sludge to the land now. San Francisco is initiating a study of such a plan.[72] There are numerous, successful land-application systems utilizing liquid or dewatered sludge throughout the U.S. which accounted for disposal of 20% of the sludges in 1972.[1] The EPA projects that this will increase to 25% in 1985.[1]

Prior to applying sludge to the land, sludge stabilization is required to avoid nuisance conditions and minimize health hazards. Digestion (anaerobic or aerobic) is the most commonly used stabilization technique (see Chapter 5 for costs of digestion).

Typically, the land-application site is remote from the plant site. The sludge may be transported by truck, barge, railroad, or pipeline, and transport costs may comprise a very significant portion of the overall costs of land application. These are discussed in Chapter 8.

Storage of sludge between treatment and land application is usually required because the application rate of the sludge to the land is usually not the same as the rate at which sludge is generated. Treated sludge will be generated at a nearly constant rate, whereas the sludge-disposal rate will depend on weather conditions, field conditions, and the method of application.

The critical factor for determining the volume of the storage facility is the length of time the disposal area cannot be used. The influences of the method of application, the climatic conditions, and the specific site may require very small storage volume or storage for several months. Where storage requirements are minimal, a second-stage anaerobic digester may be used for storage. A covered digester is well suited for sludge storage because it will contain odorous gases which may be a problem in open basins or lagoons.

There are two philosophies concerning land-application operations: (1) apply the sludge to a plot of land which will be used for growing agricultural products or other vegetation (parkland, forests, etc.); (2) dedicate the area to the disposal of sludge with no attempt to grow crops. Overall management of a system using sludge for crop growth is more complex because the needs of the crop must be carefully balanced against sludge-disposal considerations.

The advantage of an agricultural operation in conjunction with sludge disposal is the beneficial use of nutrients in the sludge and the removal of nitrogen, heavy metals, etc., from the soil. It has been estimated that some 480,000 tons of phosphorus could be recycled if land application were practiced for all sludges when secondary treatment is applied to all U.S. wastewaters.[73] The EPA recognizes this advantage, but also expresses a concern over possible human food-chain effects:

> Although utilization of sewage sludges as a resource to recover nutrients and other benefits has been encouraged by PL 92-500 and the EPA Science Advisory Board, the workgroup members and others involved in developing this Technical Bulletin have received conflicting opinions concerning the overall merits vs. hazards of applying sludges to cropland. Possible adverse effects upon the human food chain (e.g., potential for increasing human cadium intake) has remained a major concern expressed whenever this practice is considered. The relative risks of applying sewage sludges to croplands, when compared to other routes through which these contaminants enter the human diet, have yet to be determined.[74]

Sludge constituents such as viruses, organics, cysts, and parasites are of concern from the standpoint of their ultimate fate and effect on the environment. However, they do not usually limit the *rate* at which sludge is applied to the land. Those constituents of sludge which potentially may limit the application rate of the sludge to the land are the amount of water in the sludge, the amount of nitrogen in the sludge, and the quantity of heavy metals in the sludge.

If surface runoff is to be prevented, the application of water to land obviously cannot exceed the amount of water lost by percolation, evaporation, and transpiration. While not of concern with dewatered sludges, many systems apply liquid sludge to the land. The amount of water which may be applied will vary depending on the climatic conditions, the type of soil, whether vegetation grows on the disposal site, and the type of vegetation which may be grown on the disposal site. Although the water application rate to the land should be considered, for sludges having a dry-solids content greater than 2%, water content usually does not limit the rate at which sludge may be applied to the land. For example, a 2% sludge applied at a liquid application rate of 1 inch per week will result in a solids loading of 120 tons per year per acre. At this application rate, normally some other constituent (such as nitrogen) in the sludge will control the application rate.

If nitrogen pollution of the groundwater is a concern, the amount of nitrogen in the sludge may limit the annual sludge-application rate. The nitrogen concentration of sewage sludges should be measured for each sludge. For an anaerobically digested raw sludge, the total is typically 50–70 pounds of nitrogen per ton of dry sludge solids. The amount of nitrogen (as N) found in waste-activated sludges or aerobically digested sludges is generally higher than that of raw or anaerobically digested sludges, and typically ranges from 100 to 120 pounds of nitrogen per ton of the dry-weight sludge solids.

The amount of nitrogen contained in the sludge is a concern because of the potential for nitrogen to leach to the groundwater in the form of nitrate. The concentration of nitrate (as N) is limited in potable water supplies to 10 milligrams per liter. The fate of nitrogen in soils and in groundwater is difficult to predict with accuracy because of the many processes which can affect the fate of nitrogen in the soil system. There is no doubt that excessive applications of nitrogen will lead to passage of nitrogen into the groundwater. High nitrate contents are observed in Illinois and Washington in groundwaters below agricultural areas utilizing commercial fertilizers. To avoid nitrate pollution of the groundwaters, a balance between nitrogen applied in the sludge and that removed in the crop or by other mechanisms must be struck.

In a single growing season, crop uptake of nitrogen may vary from 50 to 600 pounds per acre per year depending on the specific crop growth. Typical ranges (in pounds per acre per year) are: forest crops, 20–60; field crops, 50–150; forage crops, 75–600. Consideration of the nitrogen balance may reduce the permissible sludge-loading rate from values of 100 dry tons per acre per year experienced in average conditions without concern for a nitrogen balance to as low as 5 tons per acre per year. The

Chicago, Illinois, "Prairie Plan" proposes initial application rates of 75 dry tons per acre per year to previously strip-mined land, which will taper to 20 tons of dry sludge per acre per year and an associated 1,000 pounds of nitrogen per acre per year (50 pounds N per dry ton). The sludge-application rate for the northeast water-pollution-control plant at Philadelphia, Pennsylvania, may be limited to 25 dry tons per acre per year in order to prevent nitrate-nitrogen leaching to the groundwater in excess of 10 milligrams per liter.[75] Added discussion of nitrogen balances in crop systems is presented in documents on wastewater application to the land, but is applicable to sludge systems as well.[76, 77]

The third constituent of sewage sludge which may affect the application rate of sludge to land is the heavy-metals content. Elements in sludge that are potential hazards to plants or the food chain are: boron (B), cadmium (Cd), cobalt (Co), chromium (Cr), copper (Cu), mercury (Hg), nickel (Ni), lead (Pb), and zinc (Zn). The quantity of heavy metals in sewage sludge is highly variable and depends to a great extent on the types of industry connected to the sewage-collection system and the degree of emphasis and enforcement which the operating agency imposes on limiting the heavy metals that enter the sewage-collection system. Table 7–2 presents heavy-metals concentrations for several sludges.

Among the factors that affect the toxicity of metals to plants are:

- The amount of toxic metals present in the soil.
- The specific toxic metals present: Metals differ in their toxicity to specific plants and in specific soils.
- The pH of the amended soil: The toxic-metal content safe at pH 7 can easily be lethal to most crops at pH 5.5. Land application may lead to a lowering of the soil pH due to nitrification of the NH_4–N added. Properly selected soil amendments can readily be used to overcome this potential problem.
- The organic content of the amended soil: Organic matter chelates the toxic metals and makes them less available to injure plants.
- The phosphate content of the amended soil: Phosphate is well known for reducing Zn availability to plants and decreasing the stunting injury caused by excessive levels of toxic metals.
- The cation exchange capacity (CEC): The CEC of the soil is important in binding all cations, including the toxic-metal cations. A soil with high CEC is inherently safer for disposal of sludge than a soil with low CEC.
- The plants grown on sludge-treated soil: Plant species vary widely in tolerance to heavy metals, and varieties within a species can vary three- to tenfold.

TABLE 7-2
Heavy Metals Content of Sewage Sludges (mg/kg, ppm)[a]

Metal	Range	Mean	Median
Ag, Silver	nd–960	225	90
As, Arsenic	10–50	9	8
B, Boron	200–1430	430	350
Ba, Barium	nd–3000	1460	1300
Be, Beryllium	nd	nd	nd
Cd, Cadmium	nd–1100	87	20
Co, Cobalt	nd–800	350	100
Cr, Chromium	22–30,000	1800	600
Cu, Copper	45–16,030	1250	700
Hg, Mercury	0.1–89	7	4
Mn, Manganese	100–8800	1190	400
Ni, Nickel	nd–2800	410	100
Pb, Lead	80–26,000	1940	600
Sr, Strontium	nd–2230	440	150
Se, Selenium	10–180	26	20
V, Vanadium	nd–2100	510	400
Zn, Zinc	51–28,360	3483	1800

[a] From reference 3 (nd = no data).

An EPA report presents a detailed review of the potential hazards associated with specific heavy metals.[78] The report concludes that, with correct management practices, manganese, iron, aluminum, chromium, arsenic, selenium, antimony, lead, and mercury pose relatively little hazard to crop production and plant accumulation when sludge is applied to soil. Cadmium, copper, molybdenum, nickel, and zinc can accumulate in plants and may pose a hazard to plants, animals, or humans under certain circumstances.

Cadmium is a nonessential element which can be a serious hazard to animals and humans if dietary levels are increased substantially. Cadmium's lability in soil is reduced by organic matter, clay, hydrous iron oxides, high pH, and reducing conditions. Annual cadmium-application rates, soil pH, and crop species and varieties have a major influence on the cadmium concentration in plant tissue. One study found that 35 years of sludge application resulted in large accumulations of Cd and other trace elements in the soil but no significant accumulation in the grain of corn plants, although concentrations in the leaves and roots were significantly higher than normal.[79] The following management options are available to limit cadmium accumulation in the food supply to a relatively low level on sludge-treated land: (1) maintain soil pH at or above 6.5; (2) grow crops which tend to exclude cadmium from the whole plant or from reproductive tissue; (3) apply low annual rates of cadmium, and use sludges which have a low cadmium concentration; and (4) grow nonedible crops.

Copper, although essential to plants, can become toxic to them at high concentrations. Sludges often contain appreciable levels of copper, but application of sludge to soil results in only slight to moderate increases in the copper content of plants. Under good management practices, copper in sludges will seldom be toxic to plants and should not present a hazard to the food supply.

Molybdenum is not particularly toxic to plants, even when applied at relatively high levels. As a result, molybdenum may accumulate in plants at concentrations sufficient to cause molybdenosis in ruminant animals without prior warning from plant behavior. The practice of maintaining the soil pH at 6.5 or higher results in greater solubility and availability of the molybdenum than would occur at lower pH values. Since sludges are usually very low in molybdenum, however, it is doubtful that molybdenum in sludge would present a serious hazard to the health of grazing animals except for the unusual circumstances in which forages from sites receiving high-molybdenum sludge form the major part of the animal diet.

Sludges often contain substantial quantities of nickel, which appears to be more readily available from sludges than from inorganic sources. Nevertheless, toxicity of nickel to plants occurs only on acid soils. If the soil pH is maintained at 6.5 or above, nickel should not cause toxicity to plants or pose a threat to the food supply.

Zinc, an essential element for both plants and animals, is often found in sludge at relatively high concentrations. Additions of sludge to soil may cause substantial increases in the zinc content of plants, but toxicity seldom occurs. In general, if the pH of sludge-treated soils is maintained at 6.5 or greater, zinc should not be a hazard to plants or to the food supply unless exceptionally high amounts are added in the sludge.

Heavy-metals concentrations may restrict application of sludges in the New York–New Jersey Metropolitan Area to 2 dry tons per acre per year.[80]

Sludge may be applied to the land in a variety of ways. Small plants may spread liquid sludge directly from tank trucks. In some cases, shallow trenches may be dug, filled with sludge, and covered. Sludge may also be applied through sprinkler systems using large-diameter spray nozzle

190/Handbook of Sludge-Handling Processes

openings in cases where aerosol transport can be controlled by adequate isolation of the site. In some cases, sludge has been injected into the subsoil under pressure.[81, 82] Ridge and furrow systems have also been used successfully. The method used is generally related to the quantity of sludge to be disposed of and whether crops are to be grown on the site.

The proper management of the land-application system is the key to the success of the system. The economical and technical success of the project depends on intelligent decisions, firm and established project goals, and proper monitoring of results. Monitoring of groundwater and leachate (percolate) will provide information necessary to assure protection of the groundwater. Where crops are grown, close cooperation between the treatment system management and the farming operation is required. Scheduling sludge applications with farm operations such as planting, tilling, spraying, and harvesting is vital to successful management.

It is difficult to generalize on the cost of land disposal of sludges because of the tremendous number of variables which effect cost, such as land cost, climate, soil types, distance to disposal area from treatment plant site, allowable loading rates (may range from 2 to 100 dry tons per acre per year). Chicago was experiencing costs of $70+ per dry ton in 1972 with a lengthy barge haul involved.[71] Of this total, about $20 per ton was related to the land-application portion of the project, with the remainder resulting from digestion, concentration, and transport. If the barging operation were replaced by a pipeline, total costs were projected to drop to $35 per ton. Past reports of costs at other projects range from $8 to $50 per ton.[14] A recent report estimated the costs for several alternative land-application approaches for a large city for sites 20–100 miles from the treatment plant.[60] Estimated costs were $39–$57 per ton including, in some cases, dewatering to 20% solids. Transport costs, as discussed in Chapter 8, are often the major cost item in a land-application system. Costs for the New York Metropolitan Area were estimated at $110–$185 per ton, but were adversely affected by long transport distance (100 miles), low application rates (5–10 tons per acre per year), and high land costs ($6,000 per acre).[80] Where adverse factors such as these exist, land application may not be cost effective. However, there are many municipalities where conditions are such that land application is cost effective.

Figures 7–1 through 7–4 and Table 7–3 present cost information on some components of land-application systems. Operation and maintenance requirements include expenses for the mobile equipment and monitoring requirements. A depreciation allowance is included for periodic replacement of mobile equipment.

Disposal and Land Application/191

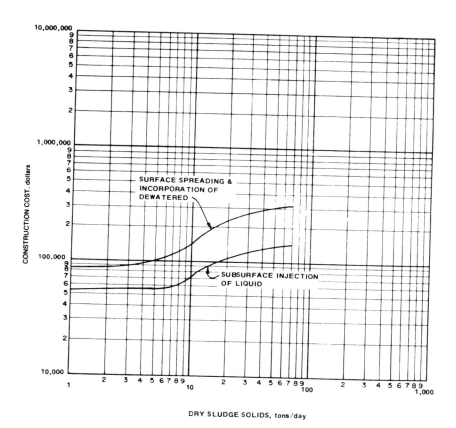

Figure 7-1. Construction costs for land disposal (Sept. 1976). Does not include land or land preparation costs.

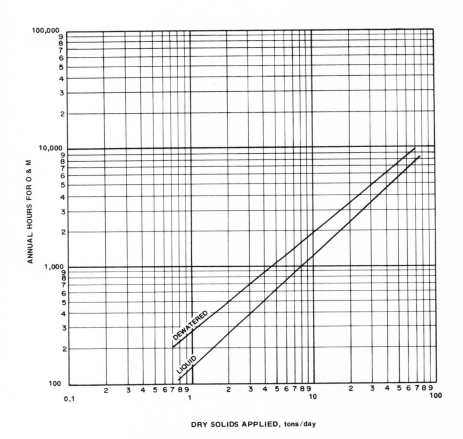

Figure 7-2. Labor requirements for land spreading of sludge.

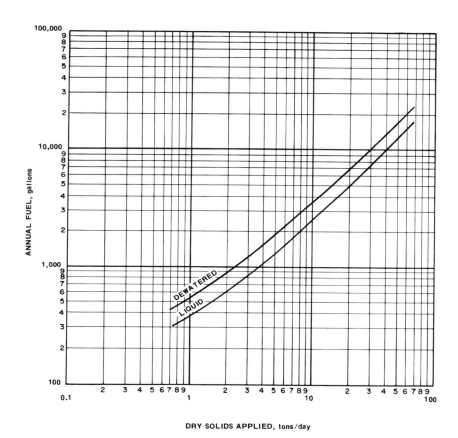

Figure 7-3. Fuel requirements for land spreading of sludge.

194 / Handbook of Sludge-Handling Processes

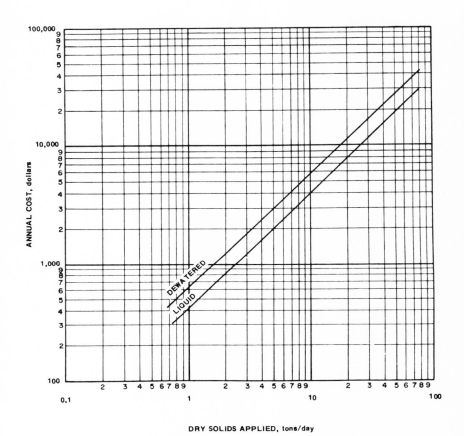

Figure 7-4. Maintenance material and supply costs for land spreading of sludge (Sept. 1976).

TABLE 7-3
Construction Costs for Land Spreading of Sludge (Sept. 1976)[a]

Cost item	Dry sludge solids (tons per day)				
	0.7	2	7	20	70
Subsurface injection of liquid sludge					
Tractor	$22,500	$22,500	$22,500	$45,500	$68,000
Injector	6,000	6,000	6,000	12,000	18,000
Hose	6,000	6,000	6,000	12,000	18,000
Housing	10,000	15,000	15,000	25,000	30,000
Contingency	6,700	7,400	7,400	14,200	20,000
Total Estimated Cost	$51,200	$56,900	$56,900	$108,700	$154,000
Surface spreading and incorporation of dewatered sludge					
Truck	$25,000	$25,000	$42,000	$84,000	$126,000
Tractor	22,000	22,000	22,000	45,500	68,000
Plow	10,000	10,000	15,000	20,000	20,000
Housing and transfer	10,000	15,000	30,000	50,000	50,000
Contingency	10,000	10,800	16,800	29,900	39,600
Total Estimated Cost	$77,000	$82,800	$125,400	$229,400	$303,600

[a] Costs do not include preparation of land, which may cost from $0 to $6,000 per acre.

Chapter 8

Sludge Transport

It is common practice to transport solids in liquid or dewatered form from one location to another as part of the treatment, disposal, or reuse steps. Significant technical and cost considerations must be evaluated in planning a transport system to achieve satisfactory results as the costs associated with transport can be very substantial.

This section will discuss general aspects of solids-transportation systems by truck, barge, railroad, and pipeline.

A very detailed sewage-sludge transport-cost study is available.[83] The purpose of this study was to develop a method of calculating transport costs for each mode using basic parameters such as gallons of fuel, operator man-hours, operating miles, and similar factors. Therefore, the information developed in the study would not grow out of date with inflation, and current unit costs could be used in making calculations at any future date. Formats are set up in the study for both manual and computer calculation of transport costs and methods of escalation. This chapter represents a general summary of the information in the study by providing typical current cost information. A copy of the complete study should be obtained if greater detail is needed.

The total costs for sludge transport consist of:

1. Point-to-point transport costs including capital and O & M.
2. Facilities capital and O & M costs (in the case of truck, barge, and railroad). Facilities are shown in Table 8–1.

The forms of sludge studied and their modes of transport are given in Table 8–2.

The most common liquid sludge concentration is 1%–4% solids, although liquid sludge up to 10% solids can be handled with relative ease. Dewatered sludges are normally 15%–20% solids and can be moved with belt conveyors or similar handling systems.

TABLE 8-1
Transportation Facilities Needed

	Transport mode		
Facility	Truck	Railroad	Barge
Liquid			
Loading storage	No[b]	Yes	Yes
Loading equipment	Yes	Yes	Yes
Dispatch office	Yes	Yes	Yes
Dock and control bldg.	N/A	N/A	Yes
Railroad siding(s)	N/A	Yes	N/A
Unloading equipment	Yes	Yes	Yes
Unloading storage[a]	No	No	No
Dewatered			
Loading storage	Yes[c]	Yes	N/A
Loading equipment	Yes	Yes	N/A
Dispatch office	Yes	Yes	N/A
Dock and control bldg.	N/A	N/A	N/A
Railroad siding(s)	N/A	Yes	N/A
Unloading equipment	Yes	Yes	N/A
Unloading storage[a]	No	No	N/A

[a] Storage assumed to be a part of another unit process.

[b] Storage required for one or two truckloads is small compared with normal plant sludge storage.

[c] Elevated storage for ease of gravity transfer to trucks. Pipeline facilities consist of pipeline and pumping stations.

TRUCK TRANSPORT

The truck is widely used for transport of both liquid and dewatered sludges. This mode offers flexibility because the terminal points and route of haul can be changed readily and at low cost. Investment in terminal facilities can be minimal. Many truck configurations are available, ranging from standard tank and dump bodies to very specialized equipment for hauling and spreading sludges. Trucks can be purchased or leased, or the hauling contracted to a private operator. The generalized costs presented are based on the following criteria and assumptions:

1. Most economical type truck from selection of standard frame or semi-trailer mounted bodies; tanks for liquid and dump, or ram type for dewatered.
2. Eight hours of trucking operation per day.
3. Fuel cost at $0.60 per gallon.

TABLE 8-2
Transport Modes for Forms of Sludge

Transport mode	Form of sludge	
	Liquid	Dewatered
Truck	X	X
Barge	X	
Railroad	X	X
Pipeline	X	

4. Amortization of truck capital cost over 6 years at 7%.
5. Truck O & M cost, excluding fuel and operator, at $0.20–$0.30 per mile, depending on type of truck.
6. Truck loading time, 30 minutes; unloading time, 15 minutes.
7. Truck average speed, 25 mph for first 20 miles one way; 35 mph for the rest.
8. General and administrative costs 25% of total O & M cost.
9. Sludge densities: liquid, 62.4 pounds per cubic foot; dewatered, 55 pounds per cubic foot; ash, 50 pounds per cubic foot.

In general, the total cost of truck transport will be decreased (per unit of material hauled) if the daily period of truck operation is increased. Restrictions may be placed on any significant truck operations such as specific routes or daylight hours for operations. The larger trucks are the most economical except for one-way haul distances less than 10 miles and annual sludge volumes less than 3,000 cubic yards for dewatered sludge and for less than 1 million gallons for liquid sludge. Generally, diesel engines are used in the larger trucks and are the economical choice for small trucks when operated at high annual mileage. Table 8-3 summarizes truck transport costs for a variety of sludge types and haul distances. Figures 8-1* through 8-4 and Table 8-4 provide supplemental cost information on truck hauling.

BARGE TRANSPORT

Barge transport has been used in the past for ocean disposal of sludges, but barges can be used for transport of sludges between land points that

*Illustrations for this chapter start on page 209.

TABLE 8-3
Truck Transport Costs

Sludge Type	% Solids	Cost,[a] (one-way haul in miles)				
		5	10	20	40	80
10 Tons Dry Solids Per Day						
Liquid	4	$35.60	$46.60	$74.00	$120.50	$200.00
	7	21.90	30.10	46.60	71.20	123.30
	11	17.50	21.40	35.60	49.30	79.50
Dewatered	20	15.90	19.70	24.10	35.60	49.30
	40	11.50	13.40	15.90	19.20	26.30
Dried	95	9.00	10.40	11.20	12.90	15.90
Ash	100	7.10	8.00	9.00	10.10	11.80
25 Tons Dry Solids Per Day						
Liquid	4	$29.60	$41.60	$68.00	$116.50	$198.20
	7	18.60	24.10	41.60	68.00	120.60
	11	14.30	18.60	28.50	47.10	77.80
Dewatered	20	10.10	14.30	19.70	28.50	48.20
	40	6.90	8.80	11.00	17.50	25.20
Dried	95	4.70	5.60	6.50	8.10	10.70
Ash	100	4.05	4.40	4.90	5.70	7.20
50 Tons Dry Solids Per Day						
Liquid	4	$26.90	$39.50	$65.80	$109.60	$191.80
	7	16.40	23.00	38.40	76.70	109.60
	11	11.50	15.30	26.30	44.90	76.70
Dewatered	20	8.80	10.40	15.90	26.90	47.10
	40	4.90	7.10	9.30	14.30	23.60
Dried	95	3.20	4.00	4.90	7.10	10.40
Ash	100	2.40	2.70	3.20	4.10	5.50
100 Tons Dry Solids Per Day						
Liquid	4	$27.40	$33.80	$54.80	$104.10	$191.80
	7	15.40	23.00	35.60	65.80	106.90
	11	10.40	15.10	24.40	46.60	74.00
Dewatered	20	6.30	8.50	14.30	26.30	46.60
	40	4.40	5.20	8.00	13.40	23.60
Dried	95	2.30	3.30	4.40	6.00	10.10
Ash	100	1.60	2.00	2.50	3.60	5.20

[a] Cost is in dollars per ton of dry solids.

TABLE 8-4
Construction Costs for Truck-Haul Facilities

Item	Annual sludge volume (cu yd)				
	1.5	5	15	50	100
Dewatered Sludge					
Conveyer	$11,000	$11,000	$11,000	$22,000	$ 22,000
Loading hopper	11,000	11,000	11,000	16,500	22,000
Loading truck encl.	5,500	5,500	5,500	11,000	11,000
Truck ramp	16,500	16,500	16,500	22,000	33,000
Unloading truck encl. and office	11,000	11,000	11,000	16,500	27,500
Total Estimated Cost	$55,000	$55,000	$55,000	$ 88,000	$115,500
Item	Annual sludge volume (million gallons)				
	1.5	5	15	50	100
Liquid Sludge					
Loading pump, pipe, hose	$ 8,200	$ 8,200	$ 9,300	$ 15,400	$ 22,000
Loading truck encl.	5,500	7,700	11,000	22,000	27,500
Truck ramp for unloading	16,500	16,500	33,000	55,000	82,000
Unloading truck encl. and office	11,000	11,000	16,500	22,000	33,000
Total Estimated Cost	$41,200	$43,400	$69,800	$114,400	$164,500

are connected by navigable waterways. The use of barges is limited to those locations in reasonable proximity to suitable waterways.

Barges have been used in the past to transport liquid sludges, but no applications for dewatered sludges are known. Barges can be leased or purchased, or the barging can be performed by an outside private operator. In most cases the towing is subcontracted to a tug operator. Self-propelled barges have been used in New York City for many years, but except for special cases, separate tugs and barges offer more flexibility.

In general, the large barges are much more cost effective than smaller barges. Larger barges have deeper drafts and, therefore, may not be practical for many inland waterways. The major factor in barging is the cost of tug (towing) services and the larger barges minimize this cost.

The information in this section is based on barges up to a 850,000-gallon capacity, but barges are available in sizes to 2 million gallons and greater. These larger sizes will substantially reduce the cost of transport for medium to large installations, but the larger barges may be too large for some inland waterways. As an example, for an annual sludge volume of 150 million gallons and a one-way haul distance of 150 miles, the total annual cost using 2-million-gallon-capacity barges would be half the total annual cost using 850,000-gallon barges.

Generalized costs were based on these criteria and assumptions:

1. Most economical barge size up to 850,000 gallons.
2. Single barge per tow.
3. Towing services contracted to outside tug operator.
4. Amortization of barge cost over 20 years at 7%.
5. Barge loading and unloading time, 5 hours each.
6. Barge average towing speed, 4 mph.
7. Barges not manned during tow.
8. General and administrative costs, 25% of total O & M cost.

Barge transit times will be variable depending on traffic, drawbridges, locks, tides, currents, and other factors. The 4-mph speed is an average, and speeds in open water may exceed 7 mph. Barges are normally unmanned during transit.

Loading can be accomplished by either a gravity pipeline or pump(s) and pipeline from a storage tank. A barge is normally filled in 2–5 hours.

Unloading requires a pump(s) for transfer of sludge to a storage system. The pump can be barge or dock mounted, and can be diesel or electric.

The use of barges was limited to liquid sludge because of the difficulty of unloading dewatered sludge from a barge and because of lack of full-scale experience.

Table 8–5 summarizes barge-transport costs.

RAILROAD TRANSPORT

It is hard to obtain information on railroad transport for generalized cases since most rail companies prefer to deal in specifics. There are very

TABLE 8-5
Barge Transport Costs

Sludge (% solids)	Cost[a] (one-way haul in miles)				
	20	40	80	160	320
10 Tons Dry Solids Per Day					
4	$82.20	$93.20	$117.80	$161.60	$268.50
7	68.50	82.20	98.60	123.30	189.00
11	63.00	74.00	87.70	106.90	139.70
25 Tons Dry Solids Per Day					
4	$48.20	$60.30	$87.70	$142.50	$230.00
7	38.40	43.80	57.00	85.50	153.40
11	32.90	37.30	48.20	64.70	93.20
50 Tons Dry Solids Per Day					
4	$34.00	$49.30	$76.70	$120.50	$219.20
7	25.80	31.80	46.60	71.20	142.50
11	21.36	25.80	35.10	50.40	93.20
100 Tons Dry Solids Per Day					
4	$27.40	$46.60	$71.20	$115.10	$216.40
7	18.90	27.40	43.80	68.50	137.00
11	14.50	19.50	30.10	46.60	90.40

[a] Cost in dollars per ton of dry solids.

few actual cases of rail transport of sludges at present, so there is little experience from which to draw information.

Rail cars can be leased from manufacturers on a full-maintenance basis. This would be the best method to assure a continuous supply of cars in good running condition. Rail companies provide a rebate of approximately $0.06–$0.20 per loaded mile (depending on condition of the car) to compensate the shipper for providing his own cars. The number of cars required is related to the round-trip transit time. Transit times have a significant effect on the number of rail cars needed and, hence, on capital or lease costs. Even with careful planning it would be difficult to reduce rail-transit time, even between close points, to less than 3 days round trip because of train makeup, switching, and weighing. Round-trip transit time typically will be 4–8 days for one-way haul distances of 20–320 miles.

Rail rates vary widely, but in general rates in various parts of the country vary according to the data in Table 8-6. Table 8-7 presents typical costs.

TABLE 8-6
Variation in Railroad Shipping Costs[a]

Area	Approximate railroad rate variation
North Central and Central	Average
Northeast	25% higher than average
Southeast	25% lower than average
Southwest	10% lower than average
West Coast	10% higher than average

[a]The following rates (per net ton) for one-way distances were used in preparing costs: 20 miles, $2.10; 40 miles, $3.00; 80 miles, $4.10; 160 miles, $6.50; and 320 miles, $12.20.

PIPELINE TRANSPORT

There are many choices to be made in the design of a sludge-pipeline system. The following assumptions were made for the purposes of this section and are representative of design criteria used in actual designs. The liquid sludge was assumed to be reasonably free of grit and grease, similar to anaerobically digested material. Raw sludge can also be transported by pipeline, but the grease may require additional maintenance procedures. The solids content does not affect the calculations within the range of 0%–4% solids. The minimum pipeline diameter is 4 inches. The literature describes installations with smaller pipelines, but these small pipelines represent special design cases.

Sludge pumps are of the dry-pit, horizontal or vertical, nonclog, centrifugal type which are widely used for sludge-pumping applications. Because of the high friction loss in 4- and 6-inch pipelines, pumping stations for these lines contain more than one pump in series in order to develop higher pumping heads and minimize the number of stations. Two pumps are operated in parallel for the 16-, 18-, and 20-inch pipelines because of the high flows. Each pumping station contains facilities for pipeline cleaning using plastic pigs and macerators.

The pipeline is cement-lined cast iron or ductile iron pipe which is typical for sludge pipelines. The cement lining provides long life and a smooth interior surface. Installation is assumed to be in normal soil conditions with average shoring and water problems typical to shallow-force main installations. Installation is assumed to be above hard rock.

TABLE 8-7
Railroad Transport Costs

Sludge Type	% Solids	Cost[a] (one-way haul in miles)				
		5	10	20	40	80
10 Tons Dry Solids Per Day						
Liquid	4	$87.70	$104.10	$134.30	$205.50	$356.20
	7	54.80	65.80	82.20	197.30	350.70
	11	43.80	52.10	65.80	85.00	137.00
Dewatered	20	25.20	35.60	41.10	49.30	82.20
	40	17.30	19.70	22.50	27.10	46.60
Dried	95	—	—	—	—	—
Ash	100	—	—	—	—	—
25 Tons Dry Solids Per Day						
Liquid	4	$84.40	$101.90	$131.50	$197.30	$350.90
	7	49.30	59.20	74.50	120.60	208.20
	11	34.00	40.60	52.60	81.10	131.50
Dewatered	20	18.60	21.90	28.50	39.50	68.00
	40	12.10	15.30	18.60	21.90	38.40
Dried	95	6.80	8.00	9.20	10.90	18.60
Ash	100	—	—	—	—	—
50 Tons Dry Solids Per Day						
Liquid	4	$82.20	$98.60	$126.00	$191.80	$339.70
	7	46.60	54.80	71.20	115.10	202.70
	11	31.20	38.40	48.20	76.70	128.80
Dewatered	20	13.20	19.20	24.10	35.60	65.80
	40	8.80	11.00	14.30	19.70	34.00
Dried	95	5.10	7.10	8.20	9.90	15.90
Ash	100	3.50	4.10	4.60	6.00	9.30
100 Tons Dry Solids Per Day						
Liquid	4	$79.50	$98.63	$123.30	$189.00	$328.80
	7	43.80	49.30	65.80	101.40	186.30
	11	30.10	37.00	46.60	74.00	126.00
Dewatered	20	12.90	18.40	23.80	34.30	63.00
	40	7.40	9.90	12.30	18.60	32.90
Dried	95	4.10	4.90	6.30	8.20	14.50
Ash	100	2.70	3.60	4.40	4.90	8.80

[a] Cost in dollars per ton of dry solids.

The pipeline cost included one major highway crossing per mile and one single-track railroad crossing per 5 miles, plus a number of driveway and several minor road crossings per mile. These costs should be typical for average installations to be expected for sludge pipelines.

The pipelines were designed based on an operating velocity of 3 feet per second. The depth of the pipeline will not affect the capital cost within the range of 3–6 feet of burial in normal soil. Facilities at the discharge end of the pipeline, such as lagoons, dewatering equipment, or spreading equipment, are assumed to be a part of other unit processes. Table 8–8 presents typical costs. Figures 8–5 through 8–7 and Table 8–9 present cost information on pipeline transport.

RELATIVE COSTS

The following general conclusions on sludge transport costs may be reached:

Liquid Sludges of 4% Solids

At a 10-tons-per-day (dry solids) capacity, pipeline is the cheapest method for distances up to 30 miles, and barge is the cheapest for greater distances.

TABLE 8–8
Pipeline Transport Costs

Sludge (% solids)	Cost[a] (distance in miles)						
	5	10	20	40	80	160	320
10 Tons Dry Solids Per Day							
0 to 4	$14.10	$28.20	$56.40	$112.80	$225.60	$451.20	$902.40
25 Tons Dry Solids Per Day							
0 to 4	$6.30	$12.60	$25.10	$50.20	$100.50	$200.90	$401.90
50 Tons Dry Solids Per Day							
0 to 4	$3.30	$6.60	$13.10	$26.30	$52.50	$105.00	$210.10
100 Tons Dry Solids Per Day							
0 to 4	$1.70	$3.50	$7.00	$13.90	$27.80	$55.60	$110.20

[a] Cost in dollars per ton of dry solids.

TABLE 8-9
Construction Costs for Pipeline Transport (Sept. 1976)

Length (ft)	Daily flow (gallons)				
	up to 100,000	250,000	500,000	750,000	1,000,000
10,000					
Pumping and terminal facilities	$ 47,000	$ 57,000	$ 71,000	$ 88,000	$108,000
Pipeline	152,000	162,500	177,500	192,500	207,500
Total Estimated Cost	$199,000	$219,500	$248,500	$280,500	$315,500
20,000					
Pumping and terminal facilities	$ 47,000	$ 57,000	$ 71,000	$ 88,000	$108,000
Pipeline	305,000	325,000	355,000	385,000	415,000
Total Estimated Cost	$352,000	$382,000	$426,000	$473,000	$523,000
50,000					
Pumping and terminal facilities	$100,000	$104,000	$ 142,000	$ 176,000	$ 216,000
Pipeline	762,500	812,500	887,500	962,500	1,037,500
Total Estimated Cost	$862,500	$916,500	$1,029,500	$1,138,500	$1,253,500
100,000					
Pumping and terminal facilities	$ 235,000	$ 178,000	$ 284,000	$ 314,300	$ 300,000
Pipeline	1,525,000	1,625,000	1,775,000	1,925,000	2,075,000
Total Estimated Cost	$1,760,000	$1,803,000	$2,059,000	$2,239,300	$2,375,000
150,000					
Pumping and terminal facilities	$ 352,500	$ 267,200	$ 426,000	$ 471,400	$ 426,300
Pipeline	2,287,500	2,437,500	2,662,500	2,887,500	3,112,500
Total Estimated Cost	$2,640,000	$2,704,700	$3,088,500	$3,358,900	$3,538,800

For 4% solids at a 25-tons-per-day (dry solids) capacity, pipeline is the choice for distances up to 55 miles, and barging at greater distances.

For plant capacities of 50 tons per day (dry solids) or more, pipeline transport is most economical for all transport distances.

Liquid Sludges of 7%-11% Solids

Truck hauling is typically lowest in cost for distances up to 20–30 miles.

For distances greater than 20 miles and for plant capacities exceeding 10 tons per day (dry solids), barge or rail transport is less expensive than trucking.

Sludge Cake and Ash

At sludge-solids concentrations greater than 20%, usually the only alternatives considered are truck and rail transport.

For plant capacities of 10 tons per day (dry solids) or less, truck transit is cheaper.

For plant capacities greater than 10 tons per day (dry solids), truck and rail transport are competitive for hauls between 20 and 80 miles; below 20 miles, trucking is cheaper; and over 80 miles, rail transit is cheaper.

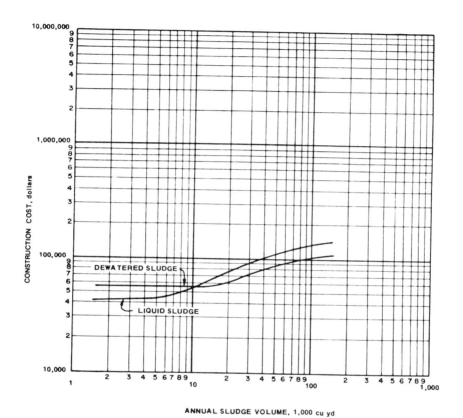

Figure 8-1. Construction costs for truck-haul facilities.

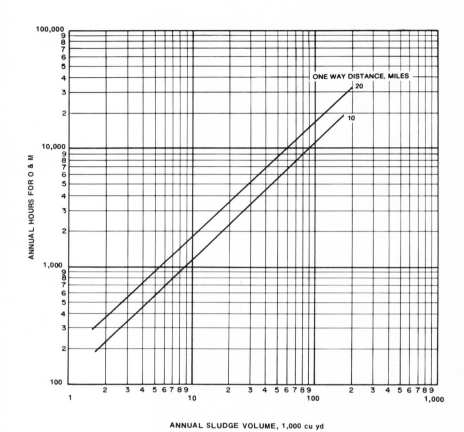

Figure 8-2. Labor requirements for sludge hauling.

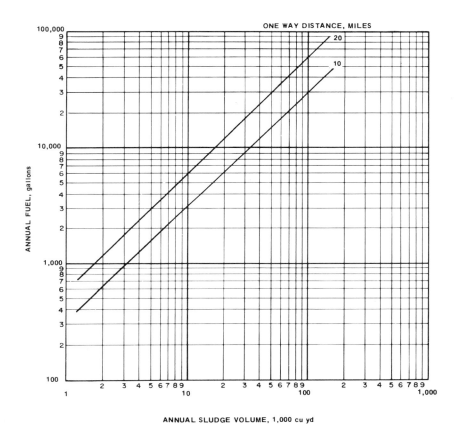

Figure 8-3. Fuel requirements for sludge hauling.

212/Handbook of Sludge-Handling Processes

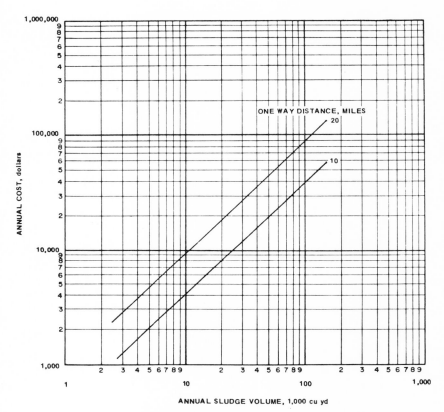

Figure 8-4. Maintenance material and supply costs for sludge hauling.

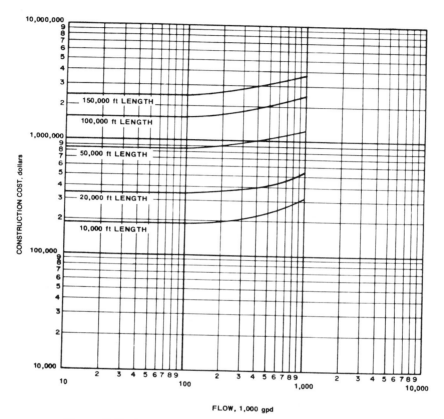

Figure 8-5. Construction costs for pipeline transport (Sept. 1976).

214/Handbook of Sludge-Handling Processes

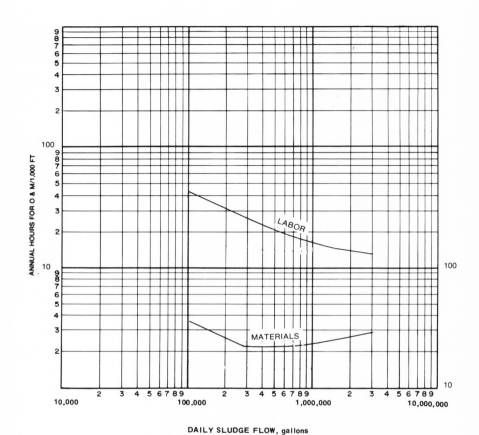

Figure 8-6. Labor requirements and maintenance material and supply costs for pipeline transport.

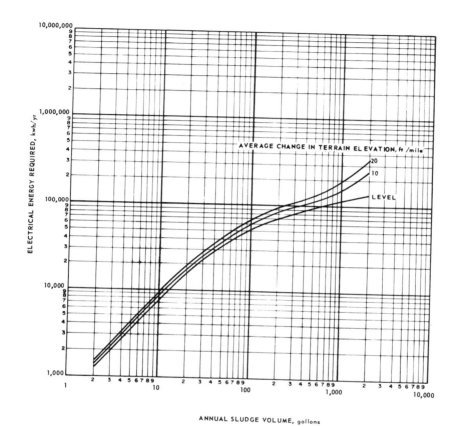

Figure 8-7. Energy requirements for pipeline transport. Design assumptions: 4% solids maximum (dilute to 4% if greater); 4-inch pipeline minimum, design velocity 3 fps; pipeline effective c factor 85; pumping based on centrifugal nonclog or slurry pumps, 68% efficiency; 20 hours per day average operation.

References

1. R. A. Olexsey, "After Ocean Disposal, What?" *Water and Wastes Engineering*, September 1976, p. 59.
2. G. L. Culp, "Environmental Pollution Control Alternatives: Municipal Wastewater," U.S. EPA Technology Transfer, EPA–625/5–76–012 (September 1976).
3. R. K. Bastian, "Municipal Sludge Management: EPA Construction Grants Program—An Overview of the Sludge Management Situation," EPA–430/9–76–009 (April 1976).
4. "Sludge Handling and Disposal Practices at Selected Municipal Wastewater Treatment Plants," EPA–430/9–77–007 (April 1977).
5. L. E. Sommers, "Chemical Composition of Sewage Sludges and Analysis of Their Potential Use as Fertilizers," Purdue University Agricultural Experiment Station Journal Paper 6420.
6. "Process Design Manual for Sludge Treatment and Disposal," U. S. EPA Technology Transfer, EPA–625/1–74–006 (October 1974).
7. "Sludge Dewatering," WPCF Manual of Practice No. 20 (1969).
8. "Utilization of Municipal Wastewater Sludge," WPCF Manual of Practice No. 2 (1971).
9. P. A. Vesilind, "Treatment and Disposal of Wastewater Sludges," Ann Arbor Science Publishers, Inc. (1974).
10. "Wastewater Treatment Plant Design," WPCF Manual of Practice No. 8 (1977).
11. C. G. Doyle, "Effectiveness of High pH for Destruction of Pathogens in Raw Sludge Filter Cake," *Journal WPCF* (1967): 1403.
12. P. A. Vesilind, "Polymer Usage Gaining for Sludge Dewatering," *Water and Wastes Engineering*, April 1971, p. 50.
13. L. J. Ewing, H. H. Almgren, and R. L. Culp, "Heat Treatment—Total Costs," paper presented at the Wastewater Treatment and Reuse Seminar, South Lake Tahoe, Calif., October 27–28, 1976.
14. R. S. Burd, "A Study of Sludge Handling and Disposal," Federal Water Pollution Control Administration Publication WP–20–4 (1968).
15. W. J. Katz and D. G. Mason, "Freezing Methods Used to Condition Activated Sludge," *Water and Sewage Works*, April 1970, p. 110.
16. "Evaluation of Conditioning and Dewatering Sewage Sludge by Freezing," EPA Report 110.0 EVE 01/71 (1971).
17. "Feasibility of Hydrolysis of Sludge Using Low Pressure Steam With SO_2

as a Hydrolytic Adjunct and Utilization of the Resulting Hydrolysate," Federal Water Pollution control Administration Report 17070 EKN 12/69 (1969).

18. R. J. Bou Thilet and R. B. Dean, "Hydrolysis of Activated Sludge," Proceedings of the 5th International Water Pollution Research Conference (1971).

19. "SO_2 Hydrolysis Converts Sludge to Animal Feed, Cuts Plant Cost," Industrial Research, October 1970, p. 31.

20. D. S. Ballantine, L. A. Miller, D. F. Bishop, and F. A. Rohrman, "The Practicality of Using Atomic Radiation for Wastewater Treatment," Journal WPCF (1967): 445.

21. J. E. Etzel, G. S. Born, J. Stein, T. J. Helbing, and G. Baney, "Sewage Sludge Conditioning and Disinfection by Gamma Irradiation," American Journal of Public Health (1967): 2067.

22. F. M. J. Compton, S. J. Black, and W. L. Whittemore, "Treating Wastewater and Sewage Sludges with Radiation: A Critical Evaluation," Nuclear News, September 1970, p. 58.

23. N. Ehlert, "Gamma Irradiation of Sewage and Sewage Sludges," Ontario Water Resources Commission, Division of Research Publication No. 38 (July, 1971).

24. R. I. Dick and K. W. Young, "Analysis of Thickening Performance of Final Settling Tanks," paper presented at the Purdue Industrial Waste Conference, May 1972.

25. R. I. Dick and B. B. Ewing, "Evaluation of Activated Sludge Thickening Theories," Journal of the Sanitary Engineering Division, ASCE, August 1967, p. 9.

26. H. J. Edde and W. W. Eckenfelder, Jr., "Theoretical Concept of Gravity Sludge Thickening: Scaling-Up Laboratory Units to Prototype Design," Journal WPCF (1969): 197.

27. A. R. Jaraheri and R. I. Dick, "Aggregate Size Variations During Thickening of Activated Sludge," Journal WPCF (1969): R197.

28. B. Fitch, "Batch Tests Predict Thickener Performance," Chemical Enigeering, August 23, 1971, p. 83.

29. W. J. Katz and A. Geinopolos, "Sludge Thickening by Dissolved Air Flotation," Journal WPCF (1967): 946.

30. W. H. Jones, "Sizing and Application of Dissolved Air Flotation Thickeners," Water and Sewage Works, Reference Issue (1968): R177.

31. G. A. Ettelt and T. J. Kennedy, "Research and Operational Experience in Sludge Dewatering at Chicago," Journal WPCF (1966): 248.

32. R. L. Braithwaite, "Polymers as Aids to the Pressure Flotation of Waste Activated Sludge," Water and Sewage Works (1964): 545.

33. S. W. Hathaway and R. A. Olexsey, "Improving Vacuum Filtration and Incineration of Sewage Sludge by Addition of Powdered Coal," Journal WPCF (1977): 2419.

34. O. E. Albertson, "Dewatering of Heat Treated Sludges," paper presented at the 42nd WPCF Conference, Dallas, Texas, October 1969.

35. C. M. Ambler, "Centrifuge Selection," Chemical Engineering, Deskbook Issue, February 15, 1971, p. 55.

36. J. R. Townsend, "What the Wastewater Plant Engineer Should Know About Centrifuges," *Water and Wastes Engineering*, November-December 1970.
37. R. V. Villiers and J. B. Farrell, "A Look at Newer Methods for Dewatering Sewage Sludges," *Civil Engineering*, December 1977, p. 66.
38. "Anaerobic Sludge Digestion," WPCF MOP No. 16 (1968).
39. "Energy Conservation in Municipal Wastewater Treatment," Culp/Wesner/Culp, EPA Contract 68-03-2186, Task 9 (1976).
40. W. F. Ettlich, "Composting as an Alternative," paper presented at the Wastewater Treatment and Reuse Seminar, South Lake Tahoe, Calif., October 1976.
41. E. Epstein, G. B. Willson, W. E. Burge, D. C. Mullen, and N. K. Enkiri, "A Forced Aeration System for Composting Sewage Sludge," *Journal WPCF* (1976): 689.
42. E. Epstein and B. G. Willson, "Composting Sewage Sludge," Biological Waste Management Laboratory, Agricultural Research Service, U.S. Department of Agriculture, Beltsville, Maryland.
43. "User Survey for Sewage Sludge Compost," Culp/Wesner/Culp, EPA Contract 68-03-2186 (May 1976).
44. M. B. Owen, "Sludge Incineration," *Journal of the Sanitary Engineering Division, ASCE* (1957): 1172.
45. J. Jacknow, "Sludge Incineration—Present State of the Art," WWEMA Ad Hoc Committee on Sludge Incineration (1975).
46. J. A. Ruf and H. T. Brown, "Generation of Electrical Energy from Municipal Refuse and Sewage Sludge," *Public Works*, January 1977, p. 38.
47. "Air Pollution Aspects of Sludge Incineration," Technology Transfer, EPA-625/4-75-009 (June 1975).
48. "Background Information for New Source Performance Standards" (Vol. 3), EPA Report 450/2-74-003, APTD-1352C (February 1974).
49. S. Balakrishman, D. E. Williamson, and R. W. Okey, "State of the Art Review on Sludge Incineration Practice," Federal Water Quality Administration Report 17070-DIV 04/70 (1970).
50. J. L. Jones, D. C. Bomberger, Jr., and F. M. Lewis, "The Economics of Energy Usage and Recovery in Sludge Disposal," paper presented at the 49th WPCF Conference, October 1976.
51. Final Report, "Sewage Sludge Incineration Task Force," EPA (February 1970).
52. L. Weller and W. Condon, "Problems in Designing Systems for Sludge Incineration," paper presented at the 16th University of Kansas Sanitary Engineering Conference, 1966.
53. D. S. Parker et al., "Lime Use in Wastewater Treatment: Design and Cost Data," EPA-600/2-75-038 (October 1975).
54. N. E. Folks, "Pyrolysis as a Means of Sewage Sludge Disposal," *Journal of the Environmental Engineering Division, ASCE*, August 1975.
55. F. M. Lewis, "Thermodynamic Fundamentals for the Pyrolysis of Refuse," Stanford Research Institute, May 1976.

56. H. W. Schultz, "Energy from Municipal Refuse: A Comparison of Ten Processes," *Professional Engineer,* November 1975.

57. N. J. Weinstein and R. F. Toro, "Thermal Processing of Municipal Solid Waste for Resource and Energy Recovery," Ann Arbor Science, Michigan (1976).

58. R. B. Sieger and B. D. Bracken, "Sludge, Garbage May Fuel California Sewage Plant," *American City and County,* January 1977, p. 37.

59. A. C. Bryan and M. T. Garrett, Jr., "What Do You Do With Sludge? Houston Has An Answer," *Public Works,* December 1972, p. 44.

60. "Innovative Technologies for Water Pollution Abatement," National Commission on Water Quality, NCWQ 75/13 (December, 1975).

61. "Basic Extractive Sludge Treatment," brochure of Resources Conservation Co., Renton, Washington (1976).

62. R. K. Ames et al., "Sludge Dewatering/Dehydration Results with MINI-B.E.S.T.," paper presented at the 30th Annual Purdue Industrial Waste Conference, Purdue University, West Lafayette, Indiana, May 1975.

63. R. K. Ames et al., "Industrial and Municipal Sludge Dewatering—The Boeing 'BEST' System," paper presented at the 29th Annual Purdue Industrial Waste Conference, West Lafayette, Indiana, May 1974.

64. R. Stone, "Practices in Disposal of Sewage Sludge by Landfill," *Public Works,* August 1972, p. 84.

65. R. Stone, "Landfill Disposal of Liquid Sewage Sludge," *Journal of the Environmental Engineering Division, ASCE,* February 1975, p. 91.

66. "Agricultural Utilization of Sewage Effluent and Sludge—An Annotated Bibliography," Federal Water Pollution Control Administration Report CWR-2 (January 1968).

67. "Land Application of Sewage Effluents and Sludges: Selected Abstracts," U.S. EPA Environmental Protection Technology Series, EPA-660/2-74-042 (June 1974).

68. *Recycling Municipal Sludges and Effluents on Land,* proceedings of a conference held in Champaign, Illinois (Washington, D. C.: National Association of State Universities and Land Grant Colleges, 1973).

69. "Recycling Treated Municipal Wastewater and Sludge Through Forest and Cropland," Pennsylvania State University Press (1973).

70. J. E. Schwing and J. L. Puntenney, "Denver Plan: Recycle Sludge to Feed Farms," *Water and Wastes Engineering,* September 1974, p. 24.

71. F. E. Dalton and R. R. Murphy, "Land Disposal IV: Reclamation and Recycle," *Journal WPCF* (1973): 1489.

72. H. C. Hyde, "Utilization of Wastewater Sludge for Agricultural Soil Enrichment," *Journal WPCF* (1976): 77.

73. T. P. Hinesley, "Sludge Recycling: The Most Reasonable Choice?" *Water Spectrum* (1973): 1.

74. "Municipal Sludge Management," EPA Technical Bulletin FRL 552-7, *Federal Register* (June 3, 1976).

75. H. H. Benjes, Jr., "Liquid Sludge Disposal on Land," paper presented at the ASCE EED Conference, Kansas City, Mo., February 1974.

76. G. M. Powell, "Land Treatment of Municipal Wastewater Effluents, Design Factors—Part II," paper presented at U.S. EPA Technology Transfer Seminars, 1975.

77. C. E. Pound and R. W. Crites, "Wastewater Treatment and Reuse By Land Application—Volumes I and II," U.S. Environmental Protection Agency Report 660/2-73-006a & b (August 1973).

78. "Application of Sewage Sludge to Cropland: Appraisal of Potential Hazards of the Heavy Metals to Plants and Animals," EPA-430/9-76-013 (November 1976).

79. M. B. Kirkham, "Trace Elements in Corn Grown on Long-Term Sludge Disposal Site," *Environmental Science and Technology,* August 1975, p. 765.

80. A. A. Kalinske, A. B. Pincince, M. H. Klegerman, and T. F. X. Flynn, "Study of Sludge Disposal Alternatives for the New York-New Jersey Metropolitan Area," paper presented at 48th WPCF Conference, October 1975.

81. C. P. Houck and J. L. Smith, "Subsurface Injection—How Much Does It Cost?" *Water and Wastes Engineering,* January 1977, p. 35.

82. J. L. Smith and C. P. Houck, "Subsurface Injection Solves Sludge Problems," *Water and Waste Engineering,* September 1976, p. 46.

83. W. F. Ettlich, "Transport of Sewage Sludge," Culp/Wesner/Culp, EPA-600/2-77-216 (December 1977).

Index

activated carbon adsorption, and odor control, 19
aerobic digestion, 120–122
　advantages, 120
　costs, 121
　disadvantages, 120
　energy requirements, 141
　labor requirements, 142
aerobic digestion system, 139
agricultural operation, in conjunction with sludge disposal, 185
air pollution, 2
　and flash drying, 161
　minimized in wet-air oxidation, 157
　national standards for incineration, 152
anaerobic digestion, 113–120
　costs, 114
　energy requirements, 130
　gas and heat available from, 118
　heat requirements
　　calculation of, 116
　　primary plus activated sludge, 133
　　primary sludge, 132
　labor requirements, 129
　two-stage, 101
animal feed, produced by sulfur dioxide hydrolysis, 21

barge transport, 199–202
　advantages, 201–202
　cost criteria, 202
　costs, 203
　disadvantages, 202
　for liquid sludge, 201

Basic Extractive Sludge Treatment (BEST), 162
　heat and electricity requirements of, 163
belt filter press, 83–84
　advantages, 83
　costs, 84, 108, 111
　energy requirements, 83, 110
　labor requirements, 109
BEST. *see* Basic Extractive Sludge Treatment
biochemical oxygen demand (BOD), 16
Blue Plains plant, Washington, D.C.
　for flash drying, 161
　operating costs, 162
BOD. *see* biochemical oxygen demand

cadmium, management options for, 188–189
centrifugal thickening. *see* thickening, centrifugal
centrifugation, 76–78
　costs, 78, 98
　energy requirements, 97
　equipment types, 76
　labor requirements, 78
　variables affecting equipment choices for, 78
　variables affecting performance, 76–78
centrifuge
　advantages of, 78
　solid-bowl, 76–77
　　performance, 77
　solid-bowl/conveyor sludge–dewatering, 76

223

224/Index

chemical conditioning, 11–16
chemical oxygen demand (COD), 16–17
chemical scrubbing, and odor control, 19
Chicago Sanitary District, and flash drying, 160
coagulants, for sewage, 5
COD. *see* chemical oxygen demand
colloidal gel systems, heat treatment of, 16
composting, 122–126
    average period, 125
    categories, 122–123
    definition, 122
    methods, 123
    solid waste, 122
    and temperature, 125
    wastewater sludge, 122
    windrow, 123–124
copper, management practices for, 189
cost components
    price indices for, 10
    of sludge management, 9–10

dewatering, 71–111
    costs, 73, 85, 88
    design criteria for, 72
    disadvantages, 72
    energy requirements for, 87
    labor requirements for, 86
    mechanical sludge removal from, 71–72
digester capacities, 117
digester gas
    cleaning and storage costs, 118–119
    storage, 117–119
    utilization, 116, 134
    utilization costs, 119–120
digester gas utilization system, 134
digester heat requirements
    primary plus activated sludge, 133
    primary sludge, 132
digestion
    aerobic, 120–122
    anaerobic, 113–120
digestion tank volume requirements, 113–114, 127
disposal alternatives, 2–3
disposal philosophy, 3

disposal plans
    integration, 3
    types of, 3
disposal plants
    integration of, 3
    types, 3
dissolved air flotation system, 58
drying beds, 71–72
    effect of climate on, 71
drying lagoons, 81–83
    costs, 83, 104, 107
    energy requirements, 106
    labor requirements, 105
drying of sludge
    flash drying, 160–162
    solvent extraction, 162–164

effluent gas cycle, in flash drying, 160
electricity generation, on-site costs, 120
environment, effect of sludge constituents on, 185
Environmental Protection Agency (EPA), 3
EPA. *see* Environmental Protection Agency
evaporation, and surface runoff, 186

ferric chloride
    construction costs for, 13, 25, 26, 27, 28
    for sludge conditioning, 11–12
fertilizer
    heat drying of sludge for, 2
    sludge processed as, 160, 161
filter cake quality, 75
filter press, 79–80, 99
    belt, 83–84
    installation, 79–80
filtration
    pressure, 79–81
    vacuum, 72–76
flash-dryer system, 180
flash drying
    cycles in, 160
    definition, 160
    disadvantages of, 161
floating aeration equipment, 121

Index/225

flotation thickening. *see* thickening, flotation
fluid-bed reactor, 176
fluid-bed reactor gases, 152
fluidized-bed furnace incineration
  energy requirements for, 177
  fuel requirements for, 178
fluidized-bed incinerator, 149, 154, 156
  supplemental fuel for, 154
freezing, 20-21
  costs, 20-21
  effect on sludge applications, 20
  and rapid dewatering, 20
furnace
  fluidized-bed, 154, 156, 157
  multiple-hearth, 153-154, 155, 156, 157

gas consumption, effects of sludge moisture and volatile solids content on, 150, 164
gases
  fluid-bed reactor, 152
  multiple-hearth incinerator, 152
gravity thickening. *see* thickening, gravity
groundwater, pollution of, 186

heat of combustion, of sludges, 150
heat exchanger, 150-151
heat treatment, 16-20
  advantage of, 17
  construction costs for, 17, 36, 37
  energy requirements for, 18, 39
  fuel requirements, 18, 38
  goal of, 16
  labor requirements for, 40
  summary of direct and indirect costs for, 20
  typical process, 17
heat-treatment-incineration system, 151
heat-treatment plants, 16-17
holding tanks
  costs, 51, 67, 70
  energy requirements, 69
  labor requirements, 68
hydrocyclone, 156
hydrolysis, with sulfur dioxide, 21

impurities, separation by centrifuges, 78
incineration, 149-160
  air pollution concerns for, 151-152
  fluidized-bed, 154, 156
  impact of excess air on, 165
  multiple-hearth, 153-154, 155, 156
  potential heat recovery from, 135, 166
  results of, 2
  supplementary fuel needed for, 151
  variables affecting cost, 150
  volume reduction by, 153
incineration, high-temperature and odor control, 19
incineration cycle, in flash drying, 160
incineration systems, major types in U.S., 149
incinerator
  fluidized-bed, 149, 154
  suspension-fired water-wall, 151
  typical multiple-hearth, 169
induced-draft cycle. *see* effluent gas cycle
industrial wastes, and wastewater sludge, 5
internal combustion engine
  costs, 135
  fuel requirements, 136
  labor requirements, 136

jet mill principle, 161-162
  sludge-drying system using, 181

lagoons. *see* drying lagoons
land application
  alternative modes for, 183-184
  costs, 190
  management of, 190
  philosophies about, 185
land disposal, construction costs for, 191
landfill operations
  costs, 183
  disadvantages of, 183
  and sewage sludge, 183
landfill site, study of, 183
lime
  construction costs for, 13, 14, 24, 32, 33, 34
  and sludge conditioning, 11-12

lime recalcining, 157
  economic feasibility of, 157
  fluidized-bed furnace for, 157
  multiple-hearth furnace for, 157
  energy requirements, 179
liquid sludges, general cost conclusions, 206, 208

mercury, EPA standards for, 152
metals
  hazards associated with, 188
  in sludge, 187, 188
  toxicity to plants, 187
methane, from anaerobic digestion, 116
Metropolitan Denver Sewage Disposal District, and flash drying, 161
molybdenum, management practices for, 189
multiple-hearth furnace, 149
  standard sizes of, *see also* multiple-hearth incinerators
multiple-hearth incineration, 153–154, 155, 156
  energy requirements for, 174
multiple-hearth incinerator gases, 152
multiple-hearth incinerators
  costs, 154, 156, 170, 175
  fuel requirements, 172
  labor requirements, 171
  start-up fuel, 173
multiple-hearth pyrolysis
  advantages, 159
  disadvantages, 159
municipal sludge incinerators, 151–152
municipal wastewater sludge, 152

nickel, toxicity of, 189
nitrate-nitrogen leaching, 187
nitrogen
  amount in sludge, 186
  crop uptake of, 186–187
nitrogen pollution, of groundwater, 186

odor control systems
  construction cost for, 19, 42
  operation and maintenance costs for, 19, 43
odors
  control of, 19, 71
    in wet-air oxidation, 157
  in drying lagoons, 82
  minimizing, 46
  produced by heat treatment, 19
organic polymeric coagulants, 11

Palo Alto incinerator, 152
partial combustion and gasification, 158
particulate collection, required efficiencies for, 152
particulate discharge requirements, 152
percolation, and surface runoff, 186
pipeline transport, 204, 206
  costs, 206, 207, 213, 214
  design criteria for, 204
  energy requirements for, 215
pollutants, gaseous, 152–153
polyelectrolytes, 48
  types of, 11
polymers
  advantages of, 11–12
  construction costs for, 13, 14, 29, 30, 31
"Prairie Plan," and nitrogen balance, 187
pressure filtration, 79–81
  costs, 81, 100, 103
  labor requirements, 81, 101
  power requirements, 81, 102
  results of, 80
primary sludge
  anaerobic digester heat requirements for, 116, 132
  anaerobic digestion of, 113
pumping sludge, 49–51
  costs, 50, 63, 66
  energy requirements, 65
  labor requirements, 64
pumps, types of, 204
pyrolysis, 157–160
  advantages of, 157–158, 159
  cost considerations, 159–160
  definition, 158
  energy production from, 159
  fuel gas produced through, 159

Index/227

v. incineration, 159
multiple-hearth
 advantages, 159
 disadvantages, 159
 as self-sustaining system, 159
 true, 158
 types, 158

quicklime, 157

radiation treatment, 21–22
 effect on sludge filterability, 22
railroad transport, 202–204
 costs, 205
  variations in, 204
 disadvantages, 203
recycling, feasibility of, 3
retention tanks, 47
reuse philosophy, of sludge handling, 2–3
rotary drum vacuum filter, 72, 89

sandbeds
 enclosure, 71
 as heat reservoir, 154, 156
sewage, chemical coagulation of, 5
sewage sludge
 BTU content of, 149
 disposal of, 4
 Sewage Sludge Incineration Task Force, 153
sludge
 activated, 1
 agricultural operations in conjunction with, 185
 characteristics, 3, 5
 chemical, 1
 combustion of, 151, 168
 digested, types of, 72
 disinfection of, 11
 disposal, 3
  in landfill, 183
 drying, 160–164
 filterability, 22
 forms and transport modes, 198, 199

heat value of, 149–150
heavy-metals content of, 187
major components of, 6–8
means of land applications, 189–190
metals present in, 152
municipal wastewater, 152
primary, 46
primary treatment of, 1
processed as fertilizer, 161
quantities from wastewater, 1
raw primary, 5
reuse, 2–3
secondary, 1, 46
secondary biological, 5
stabilization techniques, 184
storage, 184–185
transportation, 184
trends in U.S., 2
types, 2
sludge-bulking agent mixture, 123
sludge cake and ash, general cost conclusions, 208
sludge combustion, heat required for, 151, 167
sludge conditioning
 chemical, 11–16
 cost components of, 13–16
 energy required for, 12–13
 factors affecting cost of, 12
 heat treatment for, 157
 purpose of, 11
sludge-conditioning aids, 71
sludge constituents, effect on environment, 185
sludge dewatering, use of polymers, 77
sludge disposal
 cost of, 1–2
 secondary treatment, 1
sludge-handling
 alternatives, 4
 philosophic approaches to, 2–3
sludge land spreading
 costs, 194, 195
 fuel requirements, 193
 labor requirements, 192
sludge management
 operations unit for, 5–10
 significance of, 1–10

sludge-removal equipment, 71–72
sludge stabilization, 113–148
sludge transport, 197–215
  barge, 199–202, 203
  general cost conclusions, 206–208
  pipeline, 204, 206, 207
  railroad, 202–204, 205
  total cost components, 197
  truck, 198–199, 200, 201
sludge volume, effect of sludge solids on, 52
Sludge Volume Index (SVI), 48
solid waste, combustion of, 151, 168
solubilization process
  heat treatment, result of, 16
  results of, 16
solvent extraction, 162–164
  costs, 164
stack gases
  deodorization of, 150
  as potential energy source, 150–151
static-pile composting, 124–126
  costs, 145, 148
  fuel requirements, 147
  labor requirements, 146
storage facilities, volume determinations, 185
stormwater, 5
surface runoff, prevention, 185–186
SVI. *see* Sludge Volume Index

thickening
  centrifugal, 49
    WAS concentrations in, 49
  flotation, 47–49
    advantages, 47
    costs, 49, 59
    dissolved air system, 58
    labor requirements, 60
    objective, 47
    operation and performance, 48
    power requirements, 61
  gravity, 45–47
    costs, 47, 54
    labor requirements, 55
    power requirements, 56
    results of, 46

  methods, 45
  purpose of, 45
toxic substances, destruction of, 153
transpiration, and surface runoff, 186
transportation facilities, types, 198
triethylamine (TEA), in solvent extraction, 162–163
truck transport
  available types, 198
  cost criteria, 198–199
  costs, 200, 201, 209, 212
  restrictions, 199

vacuum filters, performance measurements, 73, 75
vacuum filtration, 72–76
  costs, 75, 76, 90, 93
  energy requirements, 92
  labor requirements, 75, 91
  performance measurement of, 73, 75
  and prior heat treatment, 75
Venturi scrubbers, 152
volume reduction, by sludge incineration, 153

WAS. *see* waste activated sludges
waste activated sludges (WAS), 47, 49
waste sludges, degree of thickening, 45–46
wastewater sludge, and industrial wastes, 5
wastewater treatment
  costs, 1–2
  nature of sludge resulting from, 5
  primary treatment, 1
  and sludge production, 1
water-pollution-control plant, (Philadelphia), 187
water quality, and sludge management, 1
wet-air oxidation, 16, 156–157
windrow composting, 123–124
  and climate, 123–124
  disadvantages, 125

Zimpro LPO system, 17, 35
zinc, in sludge, 189